Moscow Lectures

Volume 10

Moscow Lectures is a new book series with textbooks arising from all mathematics institutions in Moscow.

Evgeny Smirnov • Anna Tutubalina

Symmetric Functions:
A Beginner's Course

 Springer

Evgeny Smirnov
GTIIT
Shantou, Guangdong, China

Anna Tutubalina
Faculty of Mathematics
HSE University
Moscow, Russia

ISSN 2522-0314   ISSN 2522-0322   (electronic)
Moscow Lectures
ISBN 978-3-031-50340-5   ISBN 978-3-031-50341-2   (eBook)
https://doi.org/10.1007/978-3-031-50341-2

This Springer imprint is published by the registered company Springer Nature Switzerland AG
The registered company address is: Gewerbestrasse 11, 6330 Cham, Switzerland

Paper in this product is recyclable.

To the memory of
Valery Nikolaevich Tutubalin
(1936–2023)

Preface

This book is based on the notes of the undergraduate course "Symmetric functions" taught by the authors at the Mathematics department of HSE University and at the Independent University of Moscow.

Symmetric polynomials attracted the attention of mathematicians as early as the 16th century, as a tool for finding explicit formulas for solving equations of degree 3 and 4 in radicals. The fundamental theorem of symmetric polynomials, which tells us that every symmetric polynomial can be expressed as an algebraic combination of elementary symmetric polynomials, was implicitly used by Newton (although he did not state it). In the 19th century, with the development of Galois theory, symmetric polynomials became one of the key tools in algebra. In the 20th century, interest in symmetric polynomials was boosted by Issai Schur's works on representation theory. Moreover, they are related to the so-called "calculus of conditions", developed by Hermann Schubert in the late 19th century. This powerful enumerative geometry tool lacked a rigorous foundation in Schubert's time (finding one was the subject of Hilbert's 15th problem). It can be reformulated in modern terms as intersection theory on Grassmannians and flag varieties, and this is where the polynomials introduced by Schur play a crucial role, as characters of general linear groups. Conceptually these two notions are related by the famous Borel–Weil theorem.

Nowadays it is hard to find an area of fundamental mathematics that does not use symmetric polynomials: besides algebraic geometry and representation theory, they appear in mathematical physics and in topology, in probability theory and in random process theory. They are also related to numerous beautiful combinatorial constructions and problems; many of these problems still remain unsolved.

As the title suggests, this course is devoted to the theory of symmetric functions. We are mostly interested in combinatorial aspects of the theory. The course is primarily aimed at 2nd–4th year B.Sc. students majoring in mathematics; however, it can also be (and was) taken by motivated freshmen. The preliminaries required for this course are quite standard: it requires knowledge of linear algebra and some group theory and discrete mathematics (the basics of enumerative combinatorics, generating functions).

Despite the fact that the course material is intimately related to the theory of Lie groups, Lie algebras, and their representations, we have intentionally not discussed these relations in our course. This was a difficult decision, but we preferred to keep the preliminaries as modest as possible. Otherwise it would require the students to be familiar with such topics as calculus on manifolds, the basics of representation theory, and algebraic geometry. So, leaving numerous applications of symmetric functions to other areas of mathematics outside the scope of our course, we have limited ourselves to combinatorial theory, which is also a rich, interesting and often non-elementary subject. However, we hope that those students who are going to study, for example, representations of algebraic groups in finite characteristic, determinantal processes, or intersection theory on spherical varieties, will find our course useful.

This is a one-semester course consisting of one lecture and one exercise class per week, each class being 80 minutes long. The semester usually lasts from 12 to 14 weeks. The chapters of this book approximately correspond to actual lectures. Every chapter concludes with a section of problems; these or similar problems were given to students during the exercise classes.

We have tried to keep the contents of the book close to the actual class. In particular, we did not wish to mislead those instructors who are planning to use this book for their courses. This is why we have not discussed several other purely combinatorial subjects, from Macdonald polynomials to Knutson–Tao puzzles. As a partial substitute for this, we have included in each part an extra section, consisting of a series of problems; each of them covers an extra topic not appearing in the main text.

The course consists of three large parts. The first of them is devoted to Schur polynomials: in the first three chapters we give their algebraic and combinatorial definitions and prove their equivalence. In particular, this implies that the coefficients of Schur polynomials are nonnegative. Then we introduce the Hall inner product on the ring of symmetric functions, show that the Schur polynomials form an orthonormal basis for this inner product, and prove the Cauchy identity, which gives a product expansion for the sum of products of Schur polynomials in two sets of variables. Finally, we compute the principal specialization of Schur polynomials; as a consequence, we get the MacMahon formula for the number of plane partitions.

The central result of the second part is the Littlewood–Richardson rule, which provides a combinatorial interpretation of the structure constants for the basis of Schur polynomials in the ring of symmetric polynomials. To prove it, we follow an approach proposed by V. Danilov and G. Koshevoy, using the combinatorial tool of arrays. It turns out that the set of arrays bijectively corresponds to the set of pairs of semistandard Young tableaux of the same shape. This statement immediately provides another proof of the Cauchy identity. Introducing condensation operations on arrays, we get one of the simplest proofs of the Littlewood–Richardson rule.

The third part is devoted to a subject that, although not as classical as the previous ones, has remained popular among experts in both combinatorics and algebraic geometry during the last two decades: the Schubert polynomials. It is well known that Schur polynomials represent the fundamental classes of Schubert varieties in the cohomology ring of a Grassmannian; its straightforward generalization, a full flag variety, also has a basis of Schubert varieties in its cohomology ring. Schubert polynomials are "nice" representatives of these classes in the polynomial ring.

We start with preliminaries on the symmetric group and the Bruhat order. To avoid motivations from algebraic geometry, we define Schubert polynomials as "partially symmetric" polynomials, i.e. polynomials that are symmetric with respect to transpositions of certain pairs of variables. This allows us to get their initial definition, dating back to the papers of A. Lascoux and M.-P. Schützenberger [LS82] and I. N. Bernstein, I. M. Gelfand and S. I. Gelfand [BGG73]. In the last two chapters we provide a combinatorial construction for Schubert polynomials, which is parallel to the construction of Schur polynomials by means of Young tableaux; the role of the latter is played by certain configurations of pseudolines, usually called pipe dreams.

Let us point out some differences between this book and existing textbooks on the same topic. As opposed to the classical textbooks by W. Fulton [Ful97] and L. Manivel [Man98], we do not use the RSK-correspondence to construct Schur polynomials; it appears later as a byproduct of our study of arrays. Instead, we prove Littlewood's theorem on the equivalence of the algebraic and combinatorial definitions of Schur polynomials using the Lindström–Gessel–Viennot lemma on nonintersecting paths; this is one of numerous applications of this very powerful combinatorial tool. The arrays, considered in the second part, are (in our humble opinion) not as well-known as they deserve to be; the only book known to us where they are considered is the recent algebra textbook by A. Gorodentsev [Gor17]. Schubert polynomials, despite being a popular research subject, are also very rarely covered in the literature for students; the only book completely devoted to this topic, I. Macdonald's [Mac91], has been out of print for a long time and can rarely be found even in libraries. In our exposition we mostly follow A. Knutson's lecture notes [Knu12] (unpublished, but available on the author's webpage).

Let us also recommend several books for further reading on symmetric functions and their applications in various areas of mathematics. First of all, there are the books by Fulton and Manivel, mentioned above; the authors discuss symmetric functions from the point of view of the representation theory of symmetric and general linear groups, as well as their relations with Schubert calculus, i.e. intersection theory on Grassmannians and flag varieties. Another natural development of this topic leads to Macdonald polynomials; here we recommend the classical monograph by I. Macdonald [Mac98]. However, this book is written in a very compact way, so, while being an excellent reference book for experts, it can hardly be recommended to undergraduate students. Finally, let us mention a very nice book by E. Egge [Egg19]; it is based on notes of an introductory course on symmetric functions, somewhat similar to this one (however the choice of topics and approaches is quite different). It is written in a very detailed way and is recommended as a good source for a first acquaintance with the subject; on the other hand, it also covers some less standard topics, such as Grothendieck polynomials, chromatic polynomials, and Knutson–Tao puzzles.

We would like to thank all students of the *Symmetric functions* course, both at HSE University and the Independent University of Moscow, for their enthusiastic interest in the subject and for pointing out some inevitable inaccuracies and mistakes. We are deeply grateful to Grigory Merzon for numerous fruitful discussions and to Leonid Petrov for his careful reading of a draft of this book and many helpful remarks. The authors were partially supported by the HSE University Basic Research Program and by the Basis Foundation, grant "Junior Leader".

We dedicate this book to the memory of Valery Nikolaevich Tutubalin, a wonderful mathematician and A.T.'s grandfather, who passed away on June 18, 2023.

Contents

Part I
Schur Polynomials and Young Tableaux

Chapter 1
Symmetric Polynomials

Let us start with the following high school algebra problem:

Problem 1.1 Let x_1, x_2 be the roots of the quadratic equation $x^2 + px + q = 0$. Find $x_1^3 + x_2^3$.

A straightforward solution would be as follows: find the roots x_1, x_2 using the well-known formula, compute their cubes and take their sum. But this is quite a long process, and it is possible to make a mistake during the computations, so it is better to proceed in a different way: observe that, according to Viète's theorem, we have $x_1 + x_2 = -p$, $x_1 x_2 = q$. Moreover,

$$x_1^3 + x_2^3 = (x_1 + x_2)^3 - 3x_1^2 x_2 - 3x_1 x_2^2 = (x_1 + x_2)^3 - 3x_1 x_2 (x_1 + x_2) = -p^3 + 3pq.$$

This is the desired answer.

In the solution of this problem we have used the fact that the polynomial $x_1^3 + x_2^3$, as well as p and q, are *symmetric* polynomials in x_1, x_2.

1.1 Definition and First Examples

Let $R_n = \mathbb{Z}[x_1, \ldots, x_n]$ be a polynomial ring with integer coefficients in a fixed set of variables denoted by $x_1, \ldots, x_n$.

Definition 1.2 A polynomial $p(x_1, \ldots, x_n)$ is said to be *symmetric* if for any $1 \le i < n$ we have $p(\ldots, x_i, x_{i+1}, \ldots) = p(\ldots, x_{i+1}, x_i, \ldots)$.

In other words, symmetric polynomials are the invariants of the symmetric group S_n acting on R_n. Clearly, they form a subring (sums and products of symmetric polynomials are again symmetric). Let us denote it by $\Lambda_n = R_n^{S_n} \subset R_n$.

Let us give a couple of examples of symmetric polynomials:

Example 1.3 (Newton power sums) The polynomials $p_k = x_1^k + x_2^k + \cdots + x_n^k$, with $k \geq 1$, are symmetric.

Example 1.4 (Elementary symmetric polynomials) Set $e_1 = x_1 + \cdots + x_n$, $e_2 = \sum_{i<j} x_i x_j$, ..., $e_k = \sum_{i_1 < \cdots < i_k} x_{i_1} x_{i_2} \ldots x_{i_k}$, ..., $e_n = x_1 x_2 \ldots x_n$. The k-th elementary symmetric polynomial is the sum of all square-free monomials of degree k. As opposed to the previous example, k cannot exceed n. We will also formally set $e_0 = 1$ (obviously, constants are symmetric polynomials as well).

Elementary symmetric polynomials arise in Viète's formulas; these formulas relate the roots of a polynomial to its coefficients.

Theorem 1.5 (Viète) *Suppose that a polynomial $p(t) = a_0 t^n + a_1 t^{n-1} + \cdots + a_{n-1} t + a_n$ over a certain ring A has n roots $t_i \in A$:*

$$p(t) = a_0(t - t_1) \ldots (t - t_n).$$

Then $a_k = (-1)^k a_0 e_k(t_1, \ldots, t_n)$.

The proof is straightforward.

The ring R_n can be presented as a sum $R_n = \bigoplus_{k \geq 0} R_n^{(k)}$, where $R_n^{(k)}$ stands for the set of homogeneous polynomials of degree k. In simple terms this means that any polynomial can be uniquely presented as a sum of homogeneous polynomials.

This decomposition turns R_n into a *graded ring*: $R_n^{(0)} = \mathbb{Z}$ and $R_n^{(k)} \cdot R_n^{(\ell)} \subset R_n^{(k+\ell)}$.

Exercise 1.6 Show that $R_n^{(k)}$ is a free abelian group of rank $\dim R_n^{(k)} = \binom{n+k-1}{k}$.

Proposition 1.7 *The subring $\Lambda_n \subset R_n$ is a graded subring: each homogeneous component of a symmetric polynomial is symmetric.*

Proof Let p be a symmetric polynomial of degree k. Then its highest homogeneous component equals

$$p_k = \lim_{t \to \infty} p(tx_1, \ldots, tx_n)/t^k,$$

and this is again a symmetric polynomial (as a limit of symmetric polynomials). To show that the remaining components are symmetric, consider the polynomial $p - p_k$ and use induction. □

It is clear that, as opposed to the ring R_n generated by the first-degree polynomials $x_1, \ldots, x_n$, for the ring Λ_n degree-one generators are not enough: the component $\Lambda_n^{(1)}$ is one-dimensional, and, clearly, there exist (for $n \geq 2$) symmetric polynomials that are not representable as polynomials in its generator $x_1 + \cdots + x_n$.

1.2 The Fundamental Theorem of Symmetric Polynomials

One obvious way of constructing new symmetric polynomials is as follows. As we noticed before, sums and products of symmetric polynomials are again symmetric. So we can take an arbitrary *algebraic combination* of given symmetric polynomials (meaning we can take any combination of their sums and products), and it will be symmetric.

Formally, suppose polynomials $g_1, \ldots, g_r \in \Lambda_n$ are symmetric in $x_1, \ldots, x_n$, and let $f(y_1, \ldots, y_r)$ be an *arbitrary* polynomial in a "different set of variables" y_j. Then the polynomial $f(g_1, \ldots, g_r)$ obtained by *substituting* g_i into f is again symmetric (in $x_1, \ldots, x_n$).

A natural question would be to find a *system of generators* for the ring Λ_n, i.e. a set of polynomials such that every symmetric polynomial can be presented as an algebraic combination of polynomials from this set. Of course, we are interested in finding a *minimal* set of generators. Motivated by Proposition 1.7, we will be looking for a system of *homogeneous* generators.

It turns out that the elementary symmetric polynomials form such a system of generators: every symmetric polynomial can be expressed as an algebraic combination of elementary ones. Moreover, such a presentation is unique. This is the statement of the *fundamental theorem of symmetric polynomials*. In particular, uniqueness implies minimality.

Before stating this theorem, we introduce one more technical definition. For a monomial $u = a y_1^{k_1} y_2^{k_2} \ldots y_n^{k_n}$, we shall say that its *weight* is equal to

$$\mathrm{wt}\, u = k_1 + 2k_2 + 3k_3 + \cdots + nk_n.$$

The weight of a polynomial $f(y_1, \ldots, y_n)$ is defined as the largest weight of a monomial occurring in f.

The motivation for this definition is quite obvious: the weight of a polynomial $f(y_1, \ldots, y_n)$ is equal to the degree of the polynomial $f(e_1, \ldots, e_n) \in \mathbb{Z}[x_1, \ldots, x_n]$ obtained from f by substituting $y_i = e_i$.

Theorem 1.8 (The fundamental theorem of symmetric polynomials) *For every symmetric polynomial $h(x_1, \ldots, x_n)$ of degree d there exists a unique polynomial $f(y_1, \ldots, y_n)$ of weight not exceeding d such that $h(x_1, \ldots, x_n) = f(e_1, \ldots, e_n)$.*

Proof The proof is by induction over n. For $n = 1$ there is nothing to prove. Suppose the theorem is proven for polynomials in $n - 1$ variable.

Note that evaluating the k-th elementary symmetric polynomial $e_k(x_1, \ldots, x_n)$ at $x_n = 0$ yields either the k-th elementary symmetric polynomial in $n - 1$ variables; denote it by $\tilde{e}_k(x_1, \ldots, x_{n-1})$, if $k < n$, or 0, if $k = n$.

Now proceed by induction on $d = \deg h$. The base ($\deg f = 0$) is obvious. Suppose the statement holds for polynomials of degree less than d.

Take the polynomial $h(x_1, \ldots, x_n)$ and evaluate it at $x_n = 0$. The result is a *symmetric* polynomial in $x_1, \ldots, x_{n-1}$. Clearly,

$$\deg f(x_1, \ldots, x_{n-1}, 0) \leq \deg f = d.$$

By the induction hypothesis, there exists a polynomial $g(y_1, \ldots, y_{n-1})$ of weight not greater than d such that

$$h(x_1, \ldots, x_{n-1}, 0) = g(\widetilde{e}_1, \ldots, \widetilde{e}_{n-1}).$$

Now substitute e_k instead of $\widetilde{e}_k$ into g. The resulting polynomial $g(e_1, \ldots, e_{n-1})$ is again symmetric, but this time in $x_1, \ldots, x_n$. Subtract it from $h(x_1, \ldots, x_n)$:

$$h_1(x_1, \ldots, x_n) = h(x_1, \ldots, x_n) - g(e_1, \ldots, e_{n-1}).$$

This is again a symmetric polynomial of degree not greater than d (since the same is true of both summands on the right-hand side). Moreover, $h_1(x_1, \ldots, x_{n-1}, 0) = 0$. This means that h_1 is divisible by x_n. But it is symmetric, hence it is divisible by the product $x_1 \ldots x_n$, i.e. by e_n.

So $h_1 = e_n \cdot h_2(x_1, \ldots, x_n)$, where h_2 is also symmetric. Its degree satisfies $\deg h_2 = \deg h_1 - n \le d - n < d$. So we can apply the induction hypothesis: there exists a polynomial $\widehat{f}(y_1, \ldots, y_n)$ of weight not greater than $d - n$, such that $h_2(x_1, \ldots, x_n) = \widehat{f}(e_1, \ldots, e_n)$.

So we get an expression for the polynomial h:

$$\begin{aligned} h &= g(e_1, \ldots, e_{n-1}) + e_n \widehat{f}(e_1, \ldots, e_n) \\ &= \left[g(y_1, \ldots, y_{n-1}) + y_n \widehat{f}(y_1, \ldots, _n) \right]_{y_i = e_i}. \end{aligned}$$

The uniqueness of such a polynomial $g(y_1, \ldots, y_n)$ can be proven similarly. Again, proceed by induction on n. Suppose that the uniqueness does not hold, and take a nonzero polynomial of the smallest degree $g(y_1, \ldots, y_n)$ satisfying $g(e_1, \ldots, e_n) = 0$.

The polynomial g can be viewed as a polynomial in the variable y_n with coefficients from the ring $\mathbb{Z}[y_1, \ldots, y_{n-1}]$:

$$g(y_1, \ldots, y_n) = g_0(y_1, \ldots, y_{n-1}) + \cdots + g_d(y_1, \ldots, y_{n-1}) y_n^d.$$

Then $g_0 \ne 0$: indeed, otherwise g would be divisible by y_n, contradicting the minimality of the degree of g. Substituting e_i for y_i in the previous equality, we get

$$g(e_1, \ldots, e_n) = g_0(e_1, \ldots, e_{n-1}) + \cdots + g_d(e_1, \ldots, e_{n-1}) e_n^d.$$

This is an equality in the ring $\mathbb{Z}[x_1, \ldots, x_n]$. Evaluating at $x_n = 0$, we obtain that

$$g_0(\widetilde{e}_1, \ldots, \widetilde{e}_{n-1}) = 0.$$

This is a nontrivial relation among the elementary symmetric polynomials $\widetilde{e}_k$ in $n-1$ variables. Contradiction. $\square$

The fundamental theorem of symmetric polynomials can be restated in an equivalent way as follows.

Corollary 1.9 *The map*

$$\mathbb{Z}[e_1,\ldots,e_n] \to \Lambda_n$$

is a ring isomorphism. Setting $\deg e_k = k$ *turns it into a graded ring isomorphism.*

The proof of the theorem provides an algorithm for expressing arbitrary symmetric polynomials as algebraic expressions in the elementary symmetric polynomials. Alternatively, this can be done by an inductive procedure involving a lexicographic order on monomials. For details see e.g. [Vin03, § 3.8].

Exercise 1.10 Express the symmetric polynomials

$$x_1^3 + x_2^3 + x_3^3 \text{ and } x_1^4 + x_2^4 + x_3^4 - 2x_1^2x_2^2 - 2x_2^2x_3^2 - 2x_1^2x_3^2$$

in terms of e_1, e_2, e_3.

1.3 Complete Symmetric Polynomials

Definition 1.11 Define a *complete symmetric polynomial* h_k for $k > 0$ as follows:

$$h_k(x_1,\ldots,x_n) = \sum_{1 \le i_1 \le i_2 \le \cdots \le i_k \le n} x_{i_1} x_{i_2} \ldots x_{i_k}.$$

Again, we formally set $h_0 = 1$. Note that for $k > n$ the polynomials h_k are nonzero, as opposed to e_k.

Exercise 1.12 Compute the value of h_k and e_k at $x_1 = \cdots = x_n = 1$.

Exercise 1.13 Express the symmetric polynomials

$$x_1^3 + x_2^3 + x_3^3 \text{ and } x_1^4 + x_2^4 + x_3^4 - 2x_1^2x_2^2 - 2x_2^2x_3^2 - 2x_1^2x_3^2$$

in terms of h_1, h_2, h_3.

Exercise 1.14 Modify the argument used in the proof of the fundamental theorem of symmetric polynomials to show that $h_1,\ldots,h_n$ also form an algebraically independent system of generators for the ring Λ_n.

In the next section we will give a different proof of this statement: namely, we will show that each of the families e_k and h_k can be expressed in terms of the other one.

1.4 Generating Functions

Consider the generating functions

$$E(t) = \sum_{k=0}^{n} e_k(x_1,\ldots,x_n)t^k \quad \text{and} \quad H(t) = \sum_{k=0}^{\infty} h_k(x_1,\ldots,x_n)t^k.$$

Exercise 1.15 Show that

$$E(t) = \prod_{i=1}^{n}(1 + x_i t).$$

It is easy to get a similar expansion for $H(t)$:

Exercise 1.16 Show that

$$H(t) = \prod_{i=1}^{n}(1 + x_i t + x_i^2 t^2 + \cdots) = \prod_{i=1}^{n}\frac{1}{1 - x_i t}.$$

This implies the following equation on the generating functions:

$$E(t)H(-t) = 1.$$

This is an equality of two formal power series. This means that all the coefficients of the product on the left-hand side, except the constant term, are equal to zero. We obtain the following (infinite) system of equations:

$$\begin{aligned} e_1 - h_1 &= 0; \\ e_2 - e_1 h_1 + h_2 &= 0; \\ e_3 - e_2 h_1 + e_1 h_2 - h_3 &= 0; \\ &\cdots \\ e_n - e_{n-1}h_1 + e_{n-2}h_2 + \cdots + (-1)^n h_n &= 0. \end{aligned} \tag{1.1}$$

This system is triangular; it allows us to express each of the h_k, one by one, in terms of the corresponding $e_1,\ldots,e_k$. Moreover, all coefficients in this expansion will be integers.

We obtain the following corollary from the fundamental theorem of symmetric polynomials.

Corollary 1.17 *Every symmetric polynomial $p \in \Lambda_n$ can be uniquely presented as a polynomial in $h_1,\ldots,h_n$. In other words, we have a ring isomorphism*

$$\mathbb{Z}[h_1,\ldots,h_n] \to \Lambda_n.$$

Setting $\deg h_k = k$ *turns this isomorphism into a graded ring isomorphism.*

Moreover, the system of equations is symmetric with respect to interchanging e_k and h_k. This means that the coefficients of the expansion of h_k in terms of e_ℓ are the same as the coefficients occurring in the expansion of e_k in terms of h_ℓ. For example, if $h_3 = e_3 - e_2 h_1 + e_1 h_2 = e_3 - e_2 e_1 + e_1 (e_1^2 - e_2) = e_3 - 2e_1 e_2 + e_1^3$, we have $e_3 = h_3 - 2h_1 h_2 + h_1^3$. We will need this fact later, when we will be dealing with Schur polynomials.

1.5 Newton Power Sums

The ring of symmetric polynomials over $\mathbb{Q}$ (or any other field of characteristic 0), instead of $\mathbb{Z}$, possesses one more remarkable system of generators. Define the *Newton power sums* as follows:

$$p_k (x_1, \ldots, x_n) = x_1^k + \cdots + x_n^k.$$

Problem 1.2 states that elementary symmetric polynomials e_k and power sums p_k can be expressed in terms of each other. Hence the following corollary of the fundamental theorem holds:

Corollary 1.18 *Every symmetric polynomial $f \in (\mathbb{Q}[x_1, \ldots, x_n])^{S_n}$ can be uniquely expressed as a polynomial in $p_1, \ldots, p_n$. In other words, there exists a ring isomorphism*

$$\mathbb{Q}[p_1, \ldots, p_n] \to (\mathbb{Q}[x_1, \ldots, x_n])^{S_n}.$$

Setting $\deg p_k = k$ *turns it into an isomorphism of graded rings.*

Remark 1.19 It is clear that power sums do not generate the ring of symmetric polynomials with integer coefficients as a $\mathbb{Z}$-module. For example, $x_1 x_2$ cannot be expressed as an integer algebraic combination of $x_1 + x_2$ and $x_1^2 + x_2^2$ (try to find an elementary proof of this fact!).

1.6 Bases and Partitions

Let us restate the fundamental theorem of symmetric polynomials as follows. Since each symmetric polynomial $f \in \Lambda_n$ can be uniquely presented as a polynomial in $e_1, e_2, \ldots, e_n$, it can be presented as a *linear* combination of products $e_{\lambda_1} e_{\lambda_2} \cdots e_{\lambda_m}$.

Since $e_1, \ldots, e_n$ commute, we can order them: suppose that $n \geq \lambda_1 \geq \cdots \geq \lambda_m$.

Definition 1.20 A *partition* is a sequence of integers $\lambda = (\lambda_1, \ldots, \lambda_m)$ satisfying $\lambda_1 \geq \cdots \geq \lambda_m \geq 0$. The numbers λ_i are called *parts* of λ.

Thus, the polynomials

$$e_\lambda = e_{\lambda_1} \ldots e_{\lambda_m}$$

indexed by partitions λ satisfying $\lambda_1 \leq n$ form a basis in the ring of polynomials Λ_n.
Similarly we can define polynomials

$$h_\lambda = h_{\lambda_1} \ldots h_{\lambda_m}.$$

They also form a basis in Λ_n. The polynomials

$$p_\lambda = p_{\lambda_1} \ldots p_{\lambda_m}$$

do not form a $\mathbb{Z}$-basis in Λ_n, but they do form a $\mathbb{Q}$-basis of $\Lambda_n \otimes \mathbb{Q}$.

1.7 More About Partitions

When we consider products $e_{\lambda_1} \ldots e_{\lambda_m}$, we say that λ_i are integers between 1 and n. On the other hand, it can be useful to consider λ as a sequence of a certain prescribed length, possibly with some zero entries at the end of it.

We will say that adding or removing several zeros at the end of it does not change it, and will write as many zeros as we need in any given case.

There is a convenient way of presenting partitions, as *Young diagrams*.

Definition 1.21 Let $\lambda = (\lambda_1, \ldots, \lambda_m)$. Let us draw a row consisting of λ_1 boxes, another row underneath the first one consisting of λ_2 boxes, and so on (all the rows are left-adjusted). The resulting picture is called *a Young diagram* corresponding to λ.

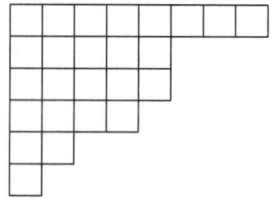

Fig. 1.1: Young diagram for the partition $\lambda = (8, 5, 5, 4, 2, 1)$

The number $\ell(\lambda)$ of nonzero elements of λ is called the *length* of λ.

The *weight* $|\lambda|$ of the partition λ is defined as the sum $\lambda_1 + \cdots + \lambda_m$ (i.e. the total number of boxes in the Young diagram). If $|\lambda| = n$, we will write it as follows: $\lambda \vdash n$, and say that λ is a *partition of n*.

Definition 1.22 If we reflect the diagram λ with respect to the diagonal, we obtain another Young diagram. This diagram corresponds to the partition λ' which is said to be *conjugate* to λ.

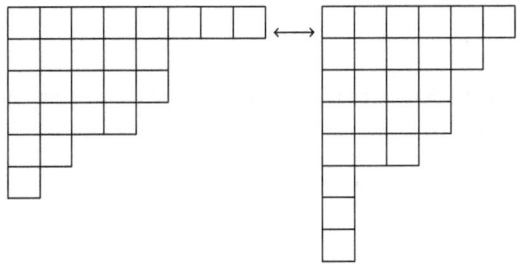

Fig. 1.2: The partitions $\lambda = (8, 5, 5, 4, 2, 1)$ and $\lambda' = (6, 5, 4, 4, 3, 1, 1, 1)$ are conjugate.

We can introduce several orderings on the set of partitions. Let us mention two of them.

Definition 1.23 (1) We shall say that $\lambda > \mu$ if λ is *greater than μ with respect to the lexicographic ordering*, i.e. for some k we have $\lambda_1 = \mu_1, \ldots, \lambda_{k-1} = \mu_{k-1}$, while $\lambda_k > \mu_k$. This is a total ordering (any two partitions are comparable).

(2) Let us introduce a *partial* ordering $\geq$ on partitions of given weight, usually called the *dominance ordering*. We shall say that $\lambda \geq \mu$ ("λ dominates μ") if the Young diagram μ is obtained from λ by moving several boxes to lower rows. Formally, this means that

$$\lambda_1 \geq \mu_1, \quad \lambda_1 + \lambda_2 \geq \mu_1 + \mu_2, \quad \lambda_1 + \lambda_2 + \lambda_3 \geq \mu_1 + \mu_2 + \mu_3, \ldots$$

Exercise 1.24 Find two partitions of the same weight that are incomparable with respect to $\geq$.

1.8 Monomial Symmetric Polynomials

We have not yet mentioned the simplest way of constructing a system of generators in Λ_n: let us take any basis in the ring of polynomials R_n and symmetrize its elements. We can do this with the basis consisting of the monomials $x_1^{i_1} \cdots x_n^{i_n}$. Let $\lambda = (\lambda_1, \ldots, \lambda_n)$ be a partition of length at most n (possibly ending with zeros). Define

$$m_\lambda = \sum_{\sigma \in S(\lambda)} x_{\sigma(1)}^{\lambda_1} \cdots x_{\sigma(n)}^{\lambda_n}.$$

The permutation σ runs over the set of all (left) cosets $S(\lambda) = S_n / \mathrm{Stab}_{S_n}(\lambda)$.

It is clear that the polynomials m_λ form a basis in Λ_n.

Remark 1.25 It would be even simpler to take the sum over all permutations from S_n, but in this case the coefficients in front of the monomials will be equal to $|\mathrm{Stab}_{S_n}(\lambda)|$, which is not necessarily equal to 1. So in this case the symmetrized polynomials will not form a basis of $(\mathbb{Z}[x_1, \ldots, x_n])^{S(n)}$ with *integer* coefficients.

1.9 Problems

1.1 Prove the following determinantal identities for elementary and complete symmetric polynomials ($k \leq n$):

$$
h_k = \det \begin{pmatrix} e_1 & 1 & 0 & \dots & 0 \\ e_2 & e_1 & 1 & \dots & 0 \\ \vdots & \vdots & \vdots & \ddots & \vdots \\ e_{k-1} & e_{k-2} & \dots & e_1 & 1 \\ e_k & e_{k-1} & \dots & e_2 & e_1 \end{pmatrix}; \quad e_k = \det \begin{pmatrix} h_1 & 1 & 0 & \dots & 0 \\ h_2 & h_1 & 1 & \dots & 0 \\ \vdots & \vdots & \vdots & \ddots & \vdots \\ h_{k-1} & h_{k-2} & \dots & h_1 & 1 \\ h_k & h_{k-1} & \dots & h_2 & h_1 \end{pmatrix}.
$$

1.2 (a) Prove the equality

$$
\frac{d}{dt} \ln E(t) = \sum_{k \geq 1} p_k\,(\mathbf{x})\,(-t)^{k-1}.
$$

(b) Using this equality, show that

$$
p_k - e_1 p_{k-1} + e_2 p_{k-2} + \cdots + (-1)^{k-1} e_{k-1} p_1 + (-1)^k k e_k = 0,
$$

where k is an arbitrary nonnegative integer.[1]

(c) Prove the following determinantal identities:

$$
p_k = \det \begin{pmatrix} e_1 & 1 & 0 & \dots & 0 \\ 2e_2 & e_1 & 1 & \dots & 0 \\ \vdots & \vdots & \vdots & \ddots & \vdots \\ (k-1)e_{k-1} & e_{k-2} & \dots & e_1 & 1 \\ ke_k & e_{k-1} & \dots & e_2 & e_1 \end{pmatrix}; \quad k! \cdot e_k = \det \begin{pmatrix} p_1 & 1 & 0 & \dots & 0 \\ p_2 & p_1 & 2 & \dots & 0 \\ \vdots & \vdots & \vdots & \ddots & \vdots \\ p_{k-1} & p_{k-2} & \dots & p_1 & k-1 \\ p_k & p_{k-1} & \dots & p_2 & p_1 \end{pmatrix}.
$$

1.3 Express the polynomial $\prod_{i=1}^{n} (1 + x_i + x_i^2)$ in terms of elementary symmetric polynomials e_k.

1.4 (a) Let $\lambda = (\lambda_1, \lambda_2, \dots, \lambda_k) \vdash m$. Show that

$$
e_\lambda = \sum_{\mu \vdash m} M_{\lambda\mu} m_\mu,
$$

where $M_{\lambda\mu}$ is the number of $k \times n$ matrices with coefficients 0 or 1, such that the sums of elements in their rows (the *row weight* of a matrix) form the sequence λ, and the column sums of elements (the *column weight* of a matrix) form the sequence μ.

(b) Show that $M_{\lambda\mu} > 0$ implies $\mu \leq \lambda'$.

(c) Show that $M_{\lambda\lambda'} = 1$.

(d) Using this formula, express $e_{(3,1,1)}$ as a linear combination of m_λ's.

---

[1] These identities were first formulated by Newton in [New07], but without a proof. The proof you are about to obtain is due to Euler [Eul50].

1.5 (a) Let $\lambda = (\lambda_1, \lambda_2, \ldots, \lambda_k) \vdash m$. Show that

$$h_\lambda = \sum_{\mu \vdash m} N_{\lambda\mu} m_\mu,$$

where $N_{\lambda\mu}$ is the number of $k \times n$ matrices with nonnegative integer coefficients, such that their row and column weights are equal to λ and μ
respectively.

(b) Find $N_{\lambda\mu}$, where $\lambda = (\lambda_1, \lambda_2, \ldots, \lambda_k) \vdash m, \mu = 1^m$

1.6 Express $e_{(3,1,1)}$ as a linear combination of h_μ's. Check the obtained identity
using the formulas from Problems 1.4 and 1.5

Chapter 2
Schur Polynomials

2.1 Skew-Symmetric Polynomials

Along with symmetric polynomials we can consider skew-symmetric polynomials. Let us find out some things about them.

Definition 2.1 A polynomial $p(x_1, \ldots, x_n) \in \mathbb{Z}[x_1, \ldots, x_n]$ is said to be *skew-symmetric* (or *alternating*) if

$$p(x_1, \ldots, x_{i+1}, x_i, \ldots, x_n) = -p(x_1, \ldots, x_n)$$

for each $1 \le i < n$.

Since every permutation can be decomposed into a product of simple transpositions (i.e. transpositions that interchange two neighboring elements), this condition is equivalent to the following one: for each permutation $w \in S_n$ we have the identity

$$p(x_{w(1)}, \ldots, x_{w(n)}) = (-1)^w p(x_1, \ldots, x_n),$$

where by $(-1)^w$ we denote the sign of w.

Denote the set of skew-symmetric polynomials in n variables (as usual, with integer coefficients) by $\mathrm{Skew}(n)$. It is clear that it is a free abelian group, i.e. it is closed under taking integer linear combinations; however, it is not a subring, since the product of two skew-symmetric polynomials is symmetric. It is a *module* over the ring Λ_n: the product of a skew-symmetric polynomial and a symmetric one is skew-symmetric.

The simplest way to obtain a skew-symmetric polynomial from an arbitrary one is by *alternation*, or *skew-symmetrization*. It is defined similarly to symmetrization:

$$\mathrm{Alt}(p) = \sum_{w \in S_n} (-1)^w (w \circ p),$$

where $w \circ p$ denotes the action of the permutation group on polynomials by means of permuting variables.

E. Smirnov, A. Tutubalina, *Symmetric Functions: A Beginner's Course*,
Moscow Lectures 10, https://doi.org/10.1007/978-3-031-50341-2_2

Exercise 2.2 Show that:

(a) if a polynomial p is symmetric with respect to a pair of variables, then $\mathrm{Alt}(p) = 0$;
(b) if two polynomials are obtained one from another by a permutation of variables: $p_1 = v \circ p_2$, then $\mathrm{Alt}(p_1) = (-1)^v \, \mathrm{Alt}(p_2)$;
(c) the alternation operator Alt is proportional to the projector onto $\mathrm{Skew}(n)$, and its image is equal to $\mathrm{Skew}(n)$. Can you compute the proportionality coefficient?

This exercise suggests a construction of a basis in $\mathrm{Skew}(n)$: let us take all the monomials $x_1^{\alpha_1} \cdots x_n^{\alpha_n}$ and alternate them; denote the resulting skew-symmetric polynomials by a_α. As shown in the exercise, we can only consider the *decreasing* sequences α_i, since two different orderings α define proportional polynomials a_α. If two variables occur in a monomial with the same power, its alternation is zero. We obtain the following proposition.

Proposition 2.3 *The polynomials $a_\alpha = \mathrm{Alt}(x_1^{\alpha_1} \ldots x_n^{\alpha_n})$ such that $\alpha_1 > \cdots > \alpha_n$ form a basis in* $\mathrm{Skew}(n)$.

The definition of alternation and the formula representing the determinant as a sum over permutations imply that a_α can we written as the following determinant:

$$
a_\alpha = \begin{vmatrix} x_1^{\alpha_1} & x_2^{\alpha_1} & \ldots & x_n^{\alpha_1} \\ x_1^{\alpha_2} & x_2^{\alpha_2} & \ldots & x_n^{\alpha_2} \\ \ldots & \ldots & \ldots & \ldots \\ x_1^{\alpha_n} & x_2^{\alpha_n} & \ldots & x_n^{\alpha_n} \end{vmatrix} = \det(x_j^{\alpha_i}).
$$

Let us consider the "smallest" sequence of decreasing indices:

$$
\delta = (n-1, n-2, \ldots, 2, 1, 0).
$$

The corresponding skew-symmetric polynomial is the well-known *Vandermonde determinant*. It decomposes into a product of linear factors.

Proposition 2.4 *We have the equality $a_\delta = \prod_{i<j}(x_i - x_j)$.*

Proof Consider the polynomial a_δ. It is skew-symmetric, so for each pair $i < j$ it vanishes at the hyperplane $x_i = x_j$. This means that a_δ is divisible by $x_i - x_j$ in the ring of polynomials $\mathbb{Z}[x_1, \ldots, x_n]$. But all these linear polynomials are coprime, so a_δ is divisible by their product, which is the right-hand side of the identity.

Next, the degrees of both sides of the equation are equal to $n(n-1)/2$. This means that their ratio is a constant. To show that this constant is equal to 1, compute the coefficients in front of an arbitrary monomial; the simplest choice is $x_1^{n-1}x_2^{n-2} \cdots x_{n-1}$. It is easy to see that this monomial occurs in both sides with coefficient 1. □

The first part of this argument holds for *any* skew-symmetric polynomial: any skew-symmetric polynomial is divisible by a_δ. It is clear that the result is a symmetric polynomial. If the initial polynomial has integer coefficients, so will the resulting symmetric polynomial. We have obtained the following result.

Proposition 2.5 *We have an isomorphism of free abelian groups*

$$
\mathrm{Skew}(n) \to \Lambda_n, \qquad p \mapsto p/a_\delta.
$$

The basis a_α of the space $\mathrm{Skew}(n)$ is mapped to a basis of Λ_n. The latter basis will be one of the main objects of our interest; its elements are called *Schur polynomials*.

2.2 Schur Polynomials

Consider a *strictly decreasing* sequence $\alpha = (\alpha_1, \ldots, \alpha_n)$, where $\alpha_1 > \alpha_2 > \cdots > \alpha_n$. Subtract from it the sequence δ componentwise; denote the result by λ. In other words, $\lambda_i = \alpha_i - n + i$. This sequence is *weakly decreasing*: $\lambda_1 \geq \lambda_2 \geq \cdots \geq \lambda_n$, i.e. it defines a partition. Let us denote this by $\alpha = \lambda + \delta$.

Definition 2.6 Let $\lambda = (\lambda_1, \lambda_2, \ldots, \lambda_n)$ be a partition (possibly with zeros at the end). The *Schur polynomial*[1] s_λ is the ratio

$$s_\lambda = a_{\lambda+\delta} / a_\delta.$$

The discussion in the previous section immediately implies the following result.

Theorem 2.7 *Schur polynomials s_λ are symmetric polynomials with integer coefficients. They form a basis in the ring Λ_n.*

Example 2.8 Let $n = 2$. Consider a partition $\lambda = (\lambda_1, \lambda_2)$ and compute the corresponding Schur polynomial.

First, $a_\delta = \begin{vmatrix} x_1 & x_2 \\ 1 & 1 \end{vmatrix} = x_1 - x_2$. Then,

$$a_{\lambda+\delta} = \begin{vmatrix} x_1^{\lambda_1+1} & x_2^{\lambda_1+1} \\ x_1^{\lambda_2} & x_2^{\lambda_2} \end{vmatrix} = (x_1 x_2)^{\lambda_2} (x_1^{\lambda_1-\lambda_2+1} - x_2^{\lambda_1-\lambda_2+1}).$$

Using the formula for the difference of n-th powers, we obtain that

$$s_\lambda = x_1^{\lambda_1} x_2^{\lambda_2} + x_1^{\lambda_1-1} x_2^{\lambda_2+1} + \cdots + x_1^{\lambda_2} x_2^{\lambda_1}.$$

Example 2.9 Let $n = 3$, $\lambda = (2, 1, 1)$. Then (check this!)

$$s_\lambda = \frac{\begin{vmatrix} x_1^4 & x_2^4 & x_3^4 \\ x_1^2 & x_2^2 & x_3^2 \\ x_1 & x_2 & x_3 \end{vmatrix}}{\begin{vmatrix} x_1^2 & x_2^2 & x_3^2 \\ x_1 & x_2 & x_3 \\ 1 & 1 & 1 \end{vmatrix}} = x_1^2 x_2 x_3 + x_1 x_2^2 x_3 + x_1 x_2 x_3^3.$$

---

[1] These polynomials were first defined by Cauchy in [Cau15]. They are named after Issai Schur, who studied their relation with representations of general linear groups (cf. [Sch01]).

Exercise 2.10 Let us write partitions as $(\lambda_1, \ldots, \lambda_n)$, with $\lambda_i > 0$, omitting the eventual "tail" of zeros. Show that:

(a) $s_{(1)} = x_1 + \cdots + x_n = e_1 = h_1$;
(b) $s_{(k)} = h_k$ (we will derive this in the next section from the Jacobi–Trudi identity; try to find a direct proof).
(c) $s_{(1^k)} = e_k$ (we denote by (1^k) the partition consisting of k parts equal to 1).

This means that the elementary and complete symmetric polynomials are particular cases of Schur polynomials.

Looking at the examples above, we can make the following observation: the coefficients of Schur polynomials are always nonnegative. This does not follow from their definition: we take two determinants, that is, polynomials with many negative coefficients, and for some reason the coefficients of their ratio are always nonnegative! It turns out that there are deep reasons for this; we will discuss them in the next chapter.

2.3 The First Jacobi–Trudi Identity

We have described several bases in the space of symmetric polynomials, including the basis of Schur polynomials s_λ. How do we find the expansion of the latter basis in other bases, in particular, the bases of elementary and complete symmetric polynomials? The answer to this question is given by the *Jacobi–Trudi identities*.[2]

Theorem 2.11 (The first Jacobi–Trudi identity) *Let $\lambda = (\lambda_1, \lambda_2, \ldots, \lambda_n)$ be a partition. Then the following identity holds:*

$$s_\lambda = \det \left(h_{\lambda_i + j - i} \right)_{i,j=1}^{n} = \det \begin{pmatrix} h_{\lambda_1} & h_{\lambda_1 + 1} & \cdots & h_{\lambda_1 + n - 1} \\ h_{\lambda_2 - 1} & h_{\lambda_2} & \cdots & h_{\lambda_2 + n - 2} \\ \vdots & \vdots & \ddots & \vdots \\ h_{\lambda_n - n + 1} & h_{\lambda_n - n + 2} & \cdots & h_{\lambda_n} \end{pmatrix}.$$

Proof Let $1 \leq \ell \leq n$. Define

$$e_j^{(\ell)} = e_j (x_1, \ldots, x_{\ell-1}, x_{\ell+1}, \ldots, x_n).$$

This is the j-th elementary symmetric polynomial in all variables except x_ℓ. According to Viète's theorem, the generating function for these polynomials (for $j < n$) equals

$$E^{(\ell)}(t) = \sum e_j^{(\ell)} t^j = \prod_{\substack{i=1 \\ i \neq \ell}}^{n} (1 + x_i t).$$

---

[2] These identities were found by Carl Gustav Jacobi in [Jac41] and later simplified by Nicola Trudi in [Tru64].

Take the product of $H(t)$ and $E^{(\ell)}(-t)$:

$$H(t)E^{(\ell)}(-t) = \left(\sum_{k\geq 0} h_k t^k\right)\left(\sum_{j=0}^{n-1} e_j^{(\ell)}(-t)^j\right)$$

$$= \prod_{i=1}^{n} \frac{1}{1-x_i t} \prod_{\substack{i=1 \\ i\neq\ell}}^{n}(1-x_i t)$$

$$= \frac{1}{1-x_l t} = 1 + x_\ell t + x_\ell^2 t^2 + \cdots .$$

Let $\alpha = (\alpha_1 > \cdots > \alpha_n)$ be a decreasing sequence. Compare the coefficients in front of t^{α_i} in both parts of the previous identity:

$$x_l^{\alpha_i} = \sum_{j=0}^{n-1} h_{\alpha_i-j}(-1)^j e_j^{(l)} = \sum_{k=1}^{n} h_{\alpha_i-n+k}(-1)^{n-k} e_{n-k}^{(l)}. \tag{2.1}$$

(Here we formally set $h_k = 0$ for $k < 0$ and $h_k = 1$ for $k = 0$.)

Consider the following matrices:

$$A_\alpha = \begin{pmatrix} x_1^{\alpha_1} & x_2^{\alpha_1} & \cdots & x_n^{\alpha_1} \\ x_1^{\alpha_2} & x_2^{\alpha_2} & \cdots & x_n^{\alpha_2} \\ \vdots & \vdots & \ddots & \vdots \\ x_1^{\alpha_n} & x_2^{\alpha_n} & \cdots & x_n^{\alpha_n} \end{pmatrix},$$

$$H_\alpha = \begin{pmatrix} h_{\alpha_1-n+1} & h_{\alpha_1-n+2} & \cdots & h_{\alpha_1} \\ h_{\alpha_2-n+1} & h_{\alpha_2-n+2} & \cdots & h_{\alpha_2} \\ \vdots & \vdots & \ddots & \vdots \\ h_{\alpha_n-n+1} & h_{\alpha_n-n+2} & \cdots & h_{\alpha_n} \end{pmatrix},$$

$$E = \begin{pmatrix} (-1)^{n-1}e_{n-1}^{(1)} & (-1)^{n-1}e_{n-1}^{(2)} & \cdots & (-1)^{n-1}e_{n-1}^{(n)} \\ (-1)^{n-2}e_{n-2}^{(1)} & (-1)^{n-2}e_{n-2}^{(2)} & \cdots & (-1)^{n-2}e_{n-2}^{(n)} \\ \vdots & \vdots & \ddots & \vdots \\ e_0^{(1)} & e_0^{(2)} & \cdots & e_0^{(n)} \end{pmatrix}.$$

The relation (2.1) implies that

$$A_\alpha = H_\alpha E. \tag{2.2}$$

Let us compute the determinant of both parts of (2.2) for the case $\alpha = \delta = (n-1,\ldots,2,1,0)$. We see that $\det A_\delta = \prod_{i<j}(x_i - x_j) = a_\delta$ is the Vandermonde determinant, and $\det H_\delta = 1$, since H_δ is an upper-triangular matrix with 1s on the principal diagonal. This implies

$$\det E = \det A_\delta = \prod_{i<j}(x_i - x_j) = a_\delta.$$

Setting $\alpha = \lambda + \delta$, where λ is the partition from the statement of the theorem, we obtain:

$$\det \begin{pmatrix} h_{\lambda_1} & h_{\lambda_1+1} & \cdots & h_{\lambda_1+n-1} \\ h_{\lambda_2-1} & h_{\lambda_2} & \cdots & h_{\lambda_2+n-2} \\ \vdots & \vdots & \ddots & \vdots \\ h_{\lambda_n-n+1} & h_{\lambda_n-n+2} & \cdots & h_{\lambda_n} \end{pmatrix} = \det H_{\lambda+\delta} = \frac{\det A_{\lambda+\delta}}{\det E} = \frac{a_{\lambda+\delta}}{a_\delta} = s_\lambda.$$

For $\lambda = (k)$ this immediately implies the following corollary announced in the previous section.

Corollary 2.12 *We have the identity* $s_{(k)} = h_k$.

For $\lambda = (1^k)$ the Jacobi–Trudi identity gives us the second determinant from Exercise 1.1; it equals e_k.

Corollary 2.13 *We have the identity* $s_{(1^k)} = e_k$.

2.4 The Pieri Formulas

Summarizing, we have defined five bases of the ring of symmetric polynomials: $m_\lambda, e_\lambda, h_\lambda, p_\lambda, s_\lambda$, all of them indexed by partitions. Three of these five bases are multiplicative in the following sense: $e_\lambda e_\mu = e_{\lambda \sqcup \mu}$, where $\lambda \sqcup \mu$ denotes the concatenation of two sequences written in the weakly decreasing order; the same holds for h_λ and p_λ.

Unfortunately, the situation with Schur polynomials is much worse. The product $s_\lambda s_\mu$ can be expanded in the basis of Schur polynomials:

$$s_\lambda s_\mu = \sum_\nu c_{\lambda\mu}^\nu s_\nu.$$

However, finding the numbers $c_{\lambda\mu}^\nu$ (the *structure constants* of this basis) can be a complicated task. We will provide a general recipe in Part II, after we develop the necessary tools. For now, we will focus on two important particular cases: when μ is a column (1^k) or a row (k). This corresponds to the case when the Schur polynomial s_μ is equal to e_k or h_k.

Theorem 2.14 (Pieri formulas) *Let* $\lambda = (\lambda_1, \ldots, \lambda_n)$ *be a partition, viewed as a Young diagram. Denote by* $\lambda \otimes 1^k$ *(respectively* $\lambda \otimes k$*) the set of Young diagrams obtained from* λ *by adding k boxes such that no two boxes belong to the same row (respectively the same column). Then the following formulas hold:*

$$s_\lambda e_k = \sum_{\mu \in \lambda \otimes 1^k} s_\mu,$$

$$s_\lambda h_k = \sum_{\mu \in \lambda \otimes k} s_\mu.$$

Proof Let us start with the first formula. Consider the polynomial

$$E(\mathbf{x}) = \sum_{k=0}^{n} e_k(\mathbf{x}) = \prod_{j=1}^{n}(1 + x_j)$$

and the product

$$a_{\lambda+\delta}(\mathbf{x}) \cdot E(\mathbf{x}) = \det\left(x_j^{\lambda_i+n-i}(1+x_j)\right)_{i,j=1}^{n}$$

$$= \sum_{\mu_1=\lambda_1}^{\lambda_1+1} \cdots \sum_{\mu_n=\lambda_n}^{\lambda_n+1} \det\left(x_j^{\mu_i+n-i}\right)_{i,j=1}^{n}$$

$$= \sum_{\mu} a_{\mu+\delta}(\mathbf{x}), \tag{2.3}$$

where the sum is taken over the partitions μ such that $\mu_i = \lambda_i$ or $\mu_i = \lambda_i + 1$ for each i, also satisfying the inequalities $\mu_1 \geq \mu_2 \geq \cdots \geq \mu_n$ (if $\mu_i < \mu_{i+1}$, they differ exactly by 1, hence $\mu_i + n - i = \mu_{i+1} + n - i - 1$, and the corresponding determinant vanishes). This means that the Young diagram μ is obtained from λ by adding one box into certain rows.

Consider the component of degree $|\lambda| + |\delta| + k$ in the equality (2.3). We obtain the following:

$$a_{\lambda+\delta} \cdot e_k = \sum_{\mu \in \lambda \otimes 1^k} a_{\mu+\delta}.$$

Dividing by the Vandermonde determinant, we get the first Pieri formula.

$$s_\lambda \cdot e_k = \sum_{\mu \in \lambda \otimes 1^k} s_\mu.$$

The proof of the second formula is similar. Consider the series

$$H(\mathbf{x}) = \sum_{k=0}^{\infty} h_k(\mathbf{x}) = \prod_{j=1}^{n}(1 - x_j)^{-1} = \prod_{j=1}^{n}\left(1 + x_j + x_j^2 + \cdots\right)$$

For simplicity denote $\lambda + \delta$ by α and consider the product

$$a_{\lambda+\delta}(\mathbf{x}) \cdot H(\mathbf{x}) = \det\left(\frac{x_j^{\alpha_i}}{1-x_j}\right)_{i,j=1}^{n} = \det\left(x_j^{\alpha_i} + x_j^{\alpha_i+1} + x_j^{\alpha_i+2} + \cdots\right)_{i,j=1}^{n}$$

$$= \det\begin{pmatrix} x_1^{\alpha_1} + x_1^{\alpha_1+1} + \cdots & x_2^{\alpha_1} + x_2^{\alpha_1+1} + \cdots & \cdots & x_n^{\alpha_1} + x_n^{\alpha_1+1} + \cdots \\ x_1^{\alpha_2} + x_1^{\alpha_2+1} + \cdots & x_2^{\alpha_2} + x_2^{\alpha_2+1} + \cdots & \cdots & x_n^{\alpha_2} + x_n^{\alpha_2+1} + \cdots \\ \vdots & \vdots & \ddots & \vdots \\ x_1^{\alpha_n} + x_1^{\alpha_n+1} + \cdots & x_2^{\alpha_n} + x_2^{\alpha_n+1} + \cdots & \cdots & x_n^{\alpha_n} + x_n^{\alpha_n+1} + \cdots \end{pmatrix} = (*)$$

Now subtract the second row from the bottom from the last one, the third row from the bottom from the second one, ... the topmost row from the second row from the top. The determinant remains the same:

$$(*) = \det \begin{pmatrix} x_1^{\alpha_1} + x_1^{\alpha_1+1} + \cdots & x_2^{\alpha_1} + x_2^{\alpha_1+1} + \cdots & \cdots & x_n^{\alpha_1} + x_n^{\alpha_1+1} + \cdots \\ x_1^{\alpha_2} + \cdots + x_1^{\alpha_1-1} & x_2^{\alpha_2} + \cdots + x_2^{\alpha_1-1} & \cdots & x_n^{\alpha_2} + \cdots + x_n^{\alpha_1-1} \\ \vdots & \vdots & \ddots & \vdots \\ x_1^{\alpha_n} + \cdots + x_1^{\alpha_{n-1}-1} & x_2^{\alpha_n} + \cdots + x_2^{\alpha_{n-1}-1} & \cdots & x_n^{\alpha_n} + \cdots + x_n^{\alpha_{n-1}-1} \end{pmatrix}$$

$$= \sum_{\beta_1=\alpha_1}^{\infty} \sum_{\beta_2=\alpha_2}^{\alpha_1-1} \cdots \sum_{\beta_n=\alpha_n}^{\alpha_{n-1}-1} \det\left(x_j^{\beta_i}\right)_{i,j=1}^n$$

$$= \sum_{\mu_1=\lambda_1}^{\infty} \sum_{\mu_2=\lambda_2}^{\lambda_1} \cdots \sum_{\mu_n=\lambda_n}^{\lambda_{n-1}} \det\left(x_j^{\mu_i+n-i}\right)_{i,j=1}^n$$

$$= \sum_{\mu} a_{\mu+\delta}.$$

Here we have

$$\beta_1 \geq \alpha_1 > \beta_2 \geq \alpha_2 > \cdots > \beta_n \geq \alpha_n.$$

The partitions λ and μ are equal to $\lambda = \alpha - \delta$ and $\mu = \beta - \delta$ respectively, so

$$\mu_1 \geq \lambda_1 \geq \mu_2 \geq \lambda_2 \geq \cdots \geq \mu_n \geq \lambda_n.$$

This means that the Young diagram μ is obtained from λ by adding k boxes to some rows in such a way that no two boxes belong to the same column.

Dividing the component of degree $|\lambda|+|\delta|+k$ in this equality by the Vandermonde determinant, we obtain the desired formula:

$$s_\lambda h_k = \sum_{\mu \in \lambda \otimes k} s_\mu.$$

Exercise 2.15 Using the Pieri formulas, express $e_{(2,2,1)} = e_2^2 e_1$ and $h_{(2,2,1)} = h_2^2 h_1$ as linear combinations of Schur polynomials.

2.5 Problems

2.1 Compute

(a) $\det\left(x_i^{\underline{j-1}}\right)_{i,j=1}^n$, where $x^{\underline{k}} = x(x-1)\cdots(x-k+1)$ is the k-th *falling factorial* of x;

(b) $\det\left(f_j(x_i)\right)_{i,j=1}^n$, where $f_j(x)$ is a polynomial with the highest term $a_j x^{j-1}$.

2.2 Prove the determinantal identities:

(a)

$$\det\left(\frac{1}{x_i + y_j}\right)^n_{i,j=1} = \frac{\prod_{1\leq i<j\leq n}(x_i - x_j)\prod_{1\leq i<j\leq n}(y_i - y_j)}{\prod^n_{i,j=1}(x_i + y_j)};$$

(b)

$$\det\left(x_j^{i-1} + x_j^{1-i}\right)^n_{i,j=1} = 2\cdot(x_1x_2\ldots x_n)^{1-n}\prod_{1\leq i<j\leq n}(x_i - x_j)(1 - x_ix_j);$$

(c) Let $x_1,\ldots,x_n, a_2\ldots a_n, b_2,\ldots, b_n$ be variables. Show that

$$\det\left((x_j + a_n)\cdots(x_j + a_{i+1})(x_j + b_i)\ldots(x_j + b_2)\right)^n_{i,j=1}$$
$$= \prod_{1\leq i<j\leq n}(x_i - x_j)\prod_{2\leq i\leq j\leq n}(b_i - a_j).$$

2.3 Compute

(a) $\det\left(i(i-1)\ldots(i-j+1)\right)^n_{i,j=1}.$

(b) $\det\left(\frac{1}{i+j-1}\right)^n_{i,j=1}.$

2.4 Show that $s_\delta = \prod_{i<j}(x_i + x_j).$

2.5 Prove the identity

$$h_r(x_1,\ldots,x_n) = \sum^n_{k=1}x_k^{n-1+r}\prod_{i\neq k}(x_k - x_i)^{-1}.$$

2.6 (a) Using the Pieri formulas, show that

$$h_k e_l = s_{(k,1^l)} + s_{(k+1,1^{l-1})}.$$

(b) Prove the following formula in two different ways: using (a) or the Jacobi–Trudi identity:

$$s_{(a+1,1^b)} = h_{a+1}e_b - h_{a+2}e_{b-1} + h_{a+3}e_{b-2} - \cdots + (-1)^{n-1}h_{a+b+1}.$$

The left-hand side is the Schur polynomial for a hook with the "arm" of length a and the "leg" of length b.

2.7 Show that the Pieri formulas provide expressions of the Schur polynomials in terms of h_k (or e_k). Conclude that the involution $\omega: h_\lambda \mapsto e_\lambda$ transforms the Schur polynomial s_λ into the Schur polynomial for the conjugate partition $s_{\lambda'}$.

Chapter 3
Combinatorial Formula for Schur Polynomials

In the previous chapter we made the following observation: in all our examples the coefficients of Schur polynomials were nonnegative integers. As we will see in this chapter, this is always true. Moreover, these coefficients express the numbers of certain combinatorial objects: the Young tableaux.

3.1 Young Tableaux

Let $\lambda = (\lambda_1, \ldots, \lambda_n)$ be a partition of length not exceeding n. Let us draw its Young diagram.

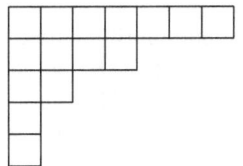

Fig. 3.1: Young diagram of shape $\lambda = (7, 4, 2, 1, 1)$.

Definition 3.1 A map from the set of boxes of a Young diagram λ into the set of integers from 1 to n is called a *semistandard Young tableau of shape* λ if it satisfies the following conditions:

- the numbers in each row weakly increase from left to right;
- the numbers in each column strictly increase from top to bottom.

Sometimes, when it is clear from the context, we will omit the word "semistandard" and call this object just a Young tableau.

Definition 3.2 The *weight* of a Young tableau T is a sequence $\mu_T = (\mu_1, \ldots, \mu_n)$, where μ_i is the number of occurrences of i in the tableau T.

E. Smirnov, A. Tutubalina, *Symmetric Functions: A Beginner's Course*,
Moscow Lectures 10, https://doi.org/10.1007/978-3-031-50341-2_3

1	1	3	4	4	4	6
2	3	5	5			
4	4					
5						
6						

Fig. 3.2: A semistandard Young tableau of shape $\lambda = (7, 4, 2, 1, 1)$ and weight $\mu = (2, 1, 2, 5, 3, 2)$

The set of all semistandard Young tableaux of shape λ filled by the numbers from 1 to n will be denoted by $\mathrm{SSYT}_\lambda(n)$.

Exercise 3.3 (a) Write down all semistandard Young tableaux of shape $\lambda = (4, 4, 3, 1)$ and weight $\mu = (4, 2, 2, 2, 2)$.
(b) Write down all tableaux from $\mathrm{SSYT}_{(2,1)}(3)$.

To each semistandard Young tableau T of weight $\mu_T = (\mu_1, \ldots, \mu_n)$ we can assign a monomial $x_1^{\mu_1} x_2^{\mu_2} \cdots x_n^{\mu_n}$. We will denote it by $\mathbf{x}^T$.

It turns out that the sum of these monomials for tableaux of given shape λ is equal to the Schur polynomial s_λ.

Theorem 3.4 (Littlewood, [Lit41]) *A Schur polynomial s_λ is equal to the sum of monomials $\mathbf{x}^T$ for all semistandard Young tableaux T of shape λ:*

$$s_\lambda(x_1, \ldots, x_n) = \sum_{T \in \mathrm{SSYT}_\lambda(n)} \mathbf{x}^T.$$

To prove this theorem, let us temporarily denote the sum $\sum_{T \in \mathrm{SSYT}_\lambda(n)} \mathbf{x}^T$ by $\widetilde{s}_\lambda$. Further in this chapter we will show that the polynomials $\widetilde{s}_\lambda$ can be computed by means of the Jacobi–Trudi identity (cf. 2.11) as a determinant with entries equal to the h_k's. Since Schur polynomials also satisfy the Jacobi–Trudi identity, this implies that $\widetilde{s}_\lambda = s_\lambda$.

Exercise 3.5 Show that Littlewood's theorem holds for Schur polynomials of a row and a column: $s_{(k)} = h_k$ and $s_{(1^k)} = e_k$ respectively.

Definition 3.6 The number of semistandard Young tableaux of shape λ and weight μ is denoted by $K_{\lambda\mu}$ and called a *Kostka number*.

Using the definition of Kostka numbers and the fact that Schur polynomials are symmetric, we can write Littlewood's theorem as follows:

$$s_\lambda = \sum_{\mu \vdash |\lambda|} K_{\lambda\mu} m_\mu.$$

Exercise 3.7 Using Littlewood's theorem, compute the coefficient in front of $x_1^4 x_2^3 x_3^3 x_4^2 x_5$ in the Schur polynomial $s_{(5,4,3,1)}(x_1, \ldots, x_5)$.

3.2 Noncrossing Paths

To show that the polynomials $\widetilde{s}_\lambda$ satisfy the Jacobi–Trudi identity, let us use another way of treating Young tableaux: as generating functions of paths on a square lattice.

Consider a $k \times (n - 1)$ rectangle on a square lattice and assign weights to the horizontal lattice segments as follows: let the segments in the bottom row be indexed by the variable x_1, the segments in the next row by x_2, ..., the segments in the top row by x_n (see Fig. 3.3). Consider a lattice path P in this rectangle joining the southwest corner with the northeast one and consisting of unit steps going right or up (the path is not allowed to go left or down). The horizontal steps of this path are indexed by variables x_i appearing in weakly increasing order. Let us call the product of all these indices the *weight* wt(P) of the path P.

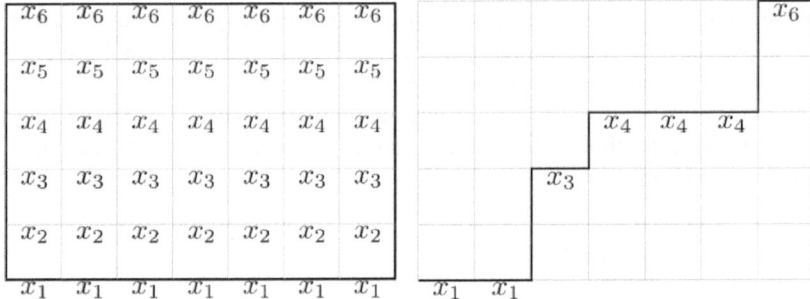

Fig. 3.3: Rectangle for $n = 6, k = 7$ and a path corresponding to the monomial $x_1^2 x_3 x_4^3 x_6$

Let us consider all possible paths in this rectangle. They correspond to all possible monomials $x_{i_1} x_{i_2} \cdots x_{i_k}$ of degree k, with $i_1 \leq i_2 \leq \cdots \leq i_k \leq n$, with each monomial occurring exactly once. We get the following proposition.

Proposition 3.8 *We have the identity*

$$\sum_{P \text{ path in the rectangle } (n-1) \times k} \text{wt}(P) = h_k(x_1, \ldots, x_n).$$

Put differently: each path corresponds to a row of length k, so it can be viewed as a one-row Young tableau filled with indices of horizontal segments (for the path in Fig. 3.3 this is $\boxed{1}\,\boxed{1}\,\boxed{3}\,\boxed{4}\,\boxed{4}\,\boxed{4}\,\boxed{6}$). Let us take a Young tableau and draw the paths corresponding to its rows on the same lattice.

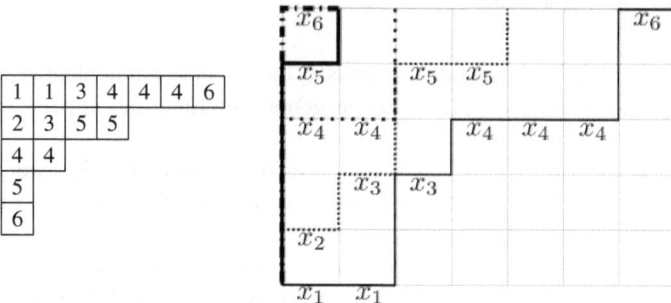

Fig. 3.4: Young tableau of shape $\lambda = (7, 4, 2, 1, 1)$ and the corresponding lattice paths

The numbers occurring in a Young tableau strictly increase along columns. This means that for two rows in the same tableau, the path corresponding to the lower row passes above the path corresponding to the higher row. The paths can have common vertical segments, but not horizontal ones.

Let us make the paths strictly non-intersecting. To do this, shift the second path to the right by 1, the third one by 2 and so on; any pair of these shifted paths will have no common points.

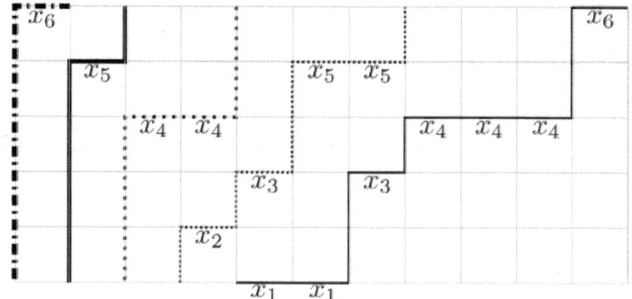

Further we will need to assign weights to collections of paths, not just to single paths. Let $\mathcal{P}$ be an arbitrary collection of paths. Denote the product of wt(P) for all paths occurring in $\mathcal{P}$ by wt($\mathcal{P}$) and call this monomial the *weight* of the collection $\mathcal{P}$.

If $\mathcal{P}$ is constructed from a Young tableau T, the weight wt($\mathcal{P}$) coincides with $\mathbf{x}^T$. This means that the sum of monomials over all Young tableaux of a given shape is equal to the sum of weights for all collections of *noncrossing* paths joining two sets of points: the starting and the terminal points of these paths.

For convenience let us introduce coordinates on our lattice (in the standard way) and provide the coordinates of these points.

Consider a Young tableau of shape $\lambda = (\lambda_1, \ldots, \lambda_m)$. It corresponds to a collection of m paths, where the i-th path starts at the point $A_i(-i; 1)$, makes λ_i steps to the right and $(n-1)$ step up and ends at the point $B_i(\lambda_i - i; n)$. The polynomial $\widetilde{s}_\lambda$ is equal to the sum

$$\widetilde{s}_\lambda = \sum_{\substack{\mathcal{P} \text{ collection of noncrossing paths} \\ A_i \to B_i}} \text{wt}(\mathcal{P}).$$

Exercise 3.9 Draw the collection of paths corresponding to the tableau

1	2	2	3	6	6
2	3	4	4		
4	4	5	6		
5	6				

3.3 The Case of Two Paths

Let us start with an easy case: compute the number of *pairs* of noncrossing paths, instead of arbitrary m-tuples. Let the first path join the points $A_1(-1; 1)$ and $B_1(\lambda_1 - 1; n)$, while the second path joins the points $A_2(-2; 1)$ and $B_2(\lambda_2 - 2; n)$.

The sum of $\text{wt}(P_1)$ over all paths $P_1 \colon A_1 \to B_1$ is equal to the polynomial h_{λ_1}. Similarly, the sum of weights $\text{wt}(P_2)$ over all paths $P_2 \colon A_2 \to B_2$ equals h_{λ_2}. This means that the sum of $\text{wt}(\mathcal{P})$ over *all* pairs of paths (without any conditions on crossings) is exactly $h_{\lambda_1} h_{\lambda_2}$.

Now let us subtract from this sum the monomials corresponding to the pairs of *intersecting* paths. To compute them, perform the following trick: find the last common point of two intersecting paths and flip the "tails" of paths after this point. Note that this transform preserves the weight $\text{wt}(\mathcal{P})$ of this pair.

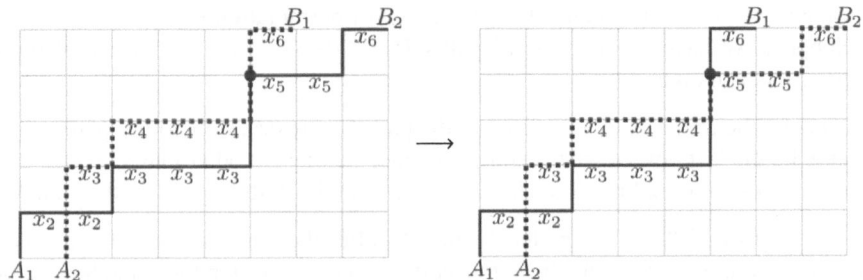

We obtain two paths joining the same points, but "in the wrong order": $A_1 \to B_2$ and $A_2 \to B_1$. This map is *involutive*: any two paths $A_1 \to B_2$ and $A_2 \to B_1$ intersect each other, and flipping their "tails" after the last intersection point gives us the initial pair of intersecting paths going from A_1 to B_1 and from A_2 to B_2.

Summarizing, we obtain a bijection between the set of pairs of intersecting paths $A_1 \to B_1, A_2 \to B_2$ and the set of *all* pairs of paths $A_1 \to B_2, A_2 \to B_1$. Moreover, this bijection is weight-preserving. The sum of weights $\text{wt}(\mathcal{P})$ over all such pairs of paths is equal to $h_{\lambda_1+1} h_{\lambda_2-1}$.

Thus, the sum of weights of *noncrossing* paths is equal to

$$\widetilde{s}_{(\lambda_1,\lambda_2)} = h_{\lambda_1} h_{\lambda_2} - h_{\lambda_1+1} h_{\lambda_2-1} = \begin{vmatrix} h_{\lambda_1} & h_{\lambda_1+1} \\ h_{\lambda_2-1} & h_{\lambda_2} \end{vmatrix}.$$

The Jacobi–Trudi formula implies that this sum is equal to $s_{(\lambda_1,\lambda_2)}$, as desired.

3.4 The General Case

Now consider the case of $m > 2$ paths. Unfortunately, we cannot repeat the previous argument: it is unclear how to count crossing paths and which pairs of tails should be flipped.

Let us proceed in a slightly different way. Consider *all* sets of paths joining the initial points with the terminal ones in an arbitrary order: $A_1 \to B_{w(1)}, \ldots, A_m \to B_{w(m)}$. Here $w \in S_m$ denotes an arbitrary permutation. Consider the sum

$$X = \sum_{\substack{w \in S_m \\ \mathcal{P} \text{ collection of paths} \\ A_1 \to B_{w(1)} \\ \cdots \\ A_m \to B_{w(m)}}} (-1)^w \operatorname{wt}(\mathcal{P}).$$

Lemma 3.10 *In the sum above, all the summands corresponding to intersecting pairs of paths occur in pairs: for each such summand, we have its negative. Hence all these summands cancel out, and the sum is equal to the sum of weights for noncrossing collections of paths.*

Proof Let us introduce an involution on the set of paths *with crossings*. This involution will preserve $\operatorname{wt}(\mathcal{P})$ and change the sign of permutation w.

Suppose we have a collection of paths $\mathcal{P} = \{P_1, \ldots, P_m\}$, and some of these paths cross each other.

- Let i_0 be the minimal index such that the path P_{i_0} crosses some other path;
- let X be the last point of the path P_{i_0} that also belongs to other paths;
- let j_0 be the minimal index such that $j_0 > i_0$ and the path P_{j_0} passes through X.

Now let us flip the "tails" of paths P_{i_0} and P_{j_0} after the point X. We get a new collection of paths $\mathcal{P}'$. The paths from this collection pass through the same horizontal segments, so $\operatorname{wt}(\mathcal{P}') = \operatorname{wt}(\mathcal{P})$. However, the new permutation w' is obtained from the initial permutation by applying a transposition $(i_0 \leftrightarrow j_0)$, so these permutations have opposite signs. $\square$

Exercise 3.11 Show that such a "flipping of tails" is indeed an involution.

Exercise 3.12 Let us try to construct an involution in a different way. Namely,

- let i_0 be the minimal index such that P_{i_0} intersects some other path;
- let j_0 be the minimal index such that P_{j_0} intersects P_{i_0};
- let X be the last intersection point of P_{i_0} and P_{j_0}.

Now let us flip the tails of P_{i_0} and P_{j_0} after the point X. Does this map indeed define an involution?

Summarizing, we have shown that X is exactly the sum we are interested in: the sum of monomials corresponding to noncrossing paths (all these paths appear with a plus sign, since for noncrossing paths $w - \mathrm{id}$). Let us compute it.

$$\widetilde{s}_\lambda = X = \sum_{\substack{w \in S_m \\ \mathcal{P} \text{ collection of paths } \substack{A_1 \to B_{w(1)} \\ \cdots \\ A_m \to B_{w(m)}}}} (-1)^w \, \mathrm{wt}(\mathcal{P})$$

$$= \sum_{w \in S_m} \sum_{P_1 : A_1 \to B_{w(1)}} \cdots \sum_{P_m : A_m \to B_{w(m)}} (-1)^w \, \mathrm{wt}(P_1) \ldots \mathrm{wt}(P_m)$$

$$= \sum_{w \in S_m} (-1)^w \left(\sum_{P_1 : A_1 \to B_{w(1)}} \mathrm{wt}(P_1) \right) \cdots \left(\sum_{P_m : A_m \to B_{w(m)}} \mathrm{wt}(P_m) \right).$$

The i-th path joins the points $A_i(-i; 1)$ and $B_{w(i)}(\lambda_{w(i)} - w(i); n)$. This means that the i-th factor is equal to $h_{\lambda_{w(i)} - w(i) + i}$.

We obtain the following identity:

$$\widetilde{s}_\lambda = X = \sum_{w \in S_m} (-1)^\sigma h_{\lambda_{w(1)} - w(1) + 1} h_{\lambda_{w(2)} - w(2) + 2} \cdots h_{\lambda_{w(m)} - w(m) + m}.$$

We see that its right-hand side is nothing but the determinant of the matrix with its (i, j)-th entry equal to $h_{\lambda_j - j + i}$,

$$\widetilde{s}_\lambda = X = \det \left(h_{\lambda_j - j + i} \right)_{i, j = 1}^m.$$

This is exactly the Jacobi–Trudi identity for s_λ. So we see that $\widetilde{s}_\lambda = s_\lambda$, as desired.

3.5 The Lindström–Gessel–Viennot Lemma

The same argument for counting noncrossing paths can be applied in many other cases. Here is the general statement.

Definition 3.13 Let $G = (V, E)$ be an oriented graph without cycles with weighted arrows: each edge $e \in E$ corresponds to its *weight* $\mathrm{wt}(e) \in R$, where R is a commutative ring. Define the *weight of the path* P to be the product of weights of its edges:

$$\mathrm{wt}(P) = \prod_{e \in P} \mathrm{wt}(e) \in R.$$

Let $\mathcal{A} = \{A_1, \ldots, A_n\} \subset V$ and $\mathcal{B} = \{B_1, \ldots, B_n\} \subset V$ be two sets of graph vertices. Consider an $n \times n$ matrix $M(\mathcal{A}, \mathcal{B}) = (m_{ij})_{i, j = 1}^n$ over the ring R such that its (i, j)-th element is equal to

$$m_{ij} = \sum_{P \text{ path from } A_i \text{ to } B_j} \text{wt}(P).$$

A *collection of paths* $\mathcal{P}$ from $\mathcal{A}$ to $\mathcal{B}$ is defined as the following data: a permutation $w \in S_n$ and an n-tuple of paths $P_i : A_i \rightarrow B_{w(i)}$. Define

$$\text{sgn}(\mathcal{P}) = \text{sgn}(w),$$

$$\text{wt}(\mathcal{P}) = \prod_{i=1}^{n} \text{wt}(P_i).$$

A collection of paths is said to be *noncrossing* if no two paths in it share a common vertex. Denote the set of all noncrossing collections from $\mathcal{A}$ to $\mathcal{B}$ by $\text{NC}(\mathcal{A}, \mathcal{B})$.

Lemma 3.14 (Lindström–Gessel–Viennot[1]) *We have the identity*

$$\det M(\mathcal{A}, \mathcal{B}) = \sum_{\mathcal{P} \in \text{NC}(\mathcal{A}, \mathcal{B})} \text{sgn}(\mathcal{P}) \, \text{wt}(\mathcal{P}).$$

Remark 3.15 In most applications of the Lindström–Gessel–Viennot lemma (or LGV lemma for short), any collection of noncrossing paths $\mathcal{P}$ joins A_i with B_i for each i, which means that $\text{sgn}(\mathcal{P}) = 1$. In this case the statement of the lemma can be rewritten as follows:

$$\det M(\mathcal{A}, \mathcal{B}) = \sum_{\mathcal{P} \in \text{NC}(\mathcal{A}, \mathcal{B})} \text{wt}(\mathcal{P}).$$

However, it is possible that noncrossing collections of paths correspond to different permutations. For example, for the initial and terminal points in the figure below, we have noncrossing collections of paths joining $A_1 \rightarrow B_1, A_2 \rightarrow B_2$, as well as those joining $A_1 \rightarrow B_2, A_2 \rightarrow B_1$.

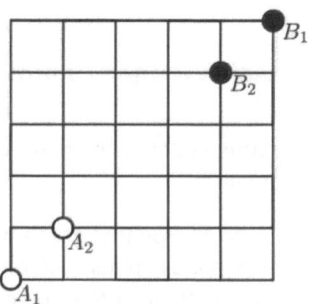

---

[1] This approach for counting collections of noncrossing paths was used by Bernt Lindström in [Lin73]. It became widely known to combinatorialists after Ira Gessel and Xavier Viennot applied it to problems of enumerative combinatorics in [GV85]. However, it appeared in probability theory much earlier, at least as early as 1959, in the paper [KM59].

Remark 3.16 If we need to compute just the *number* of noncrossing collections, instead of a generating function, we can assign weights $\mathrm{wt}(e) = 1$ to all edges $e \in E$.

3.6 Problems

3.1 Compute the Kostka numbers $K_{\lambda\mu}$ for

(a) $\lambda = (k, k)$, $\mu = (k - 1, 1^{k+1})$;
(b) $\lambda = (a, 1^b)$, $\mu = (c, 1^{a+b-c})$, where $c \leq a$;
(c) $\lambda = (k, k)$, $\mu = (1^{2k})$.

3.2 Show that $s_\lambda = m_\lambda$ plus a sum of certain monomials m_μ, where $|\mu| = |\lambda|, \mu \leq \lambda$. Use this to show that the s_λ form a basis in the ring of symmetric polynomials.

3.3 Let $\lambda = (\lambda_1, \lambda_2)$, $|\mu| = |\lambda|, \mu \leq \lambda$. Show that $K_{\lambda\mu} > 0$.

3.4 Using Pieri formulas, show that

$$h_\lambda = \sum_\mu K_{\mu\lambda} s_\mu;$$

$$e_\lambda = \sum_\mu K_{\mu\lambda} s_{\mu'}.$$

3.5 (a) Assign weights to horizontal edges of a $k \times (n - k)$ lattice in such a way that the sum of weights for the paths joining the southwest corner with the northeast one would be equal to $e_k(x_1, \ldots, x_n)$.
(b) Prove the *second Jacobi–Trudi identity* using the LGV lemma:

$$s_\lambda = \det \left(e_{\lambda'_i + j - i} \right)_{i,j=1}^n = \det \begin{pmatrix} e_{\lambda'_1} & e_{\lambda'_1+1} & \cdots & e_{\lambda'_1+n-1} \\ e_{\lambda'_2-1} & e_{\lambda'_2} & \cdots & e_{\lambda'_2+n-2} \\ \vdots & \vdots & \ddots & \vdots \\ e_{\lambda'_n-n+1} & e_{\lambda'_i-n+2} & \cdots & e_{\lambda'_n} \end{pmatrix}.$$

3.6 The n-th *Catalan number* C_n is defined as the number of paths on $\mathbb{Z}^2$, consisting of steps $(i, j) \to (i + 1, j + 1)$ and $(i, j) \to (i + 1, j - 1)$, joining the points $(-n, 0)$ and $(n, 0)$ and not passing under the X-axis. Compute the determinant

$$\det \begin{pmatrix} C_1 & C_2 & \cdots & C_n \\ C_2 & C_3 & \cdots & C_{n+1} \\ \vdots & \vdots & \ddots & \vdots \\ C_n & C_{n+1} & \cdots & C_{2n-1} \end{pmatrix}.$$

3.7 (a) Take $n+1$ distinct points $M_i(a_i, b_i) \in \mathbb{Z}^2$, with

$$0 = a_0 \le a_1 \le a_2 \le \ldots \le a_n = r \text{ and}$$
$$0 = b_0 \le b_1 \le b_2 \le \ldots \le b_n = s.$$

Show that the number of paths that consist of steps going up or right, join $(0,0)$ with (r, s) and do not pass through M_i for $i = 1, \ldots, n-1$, is equal (up to the sign) to the determinant

$$(-1)^{n-1} \det \left(\binom{a_j + b_j - a_{i-1} - b_{i-1}}{a_j - a_{i-1}} \right)_{i,j=1}^{n} =$$

$$(-1)^{n-1} \det \begin{pmatrix} \binom{a_1+b_1-a_0-b_0}{a_1-a_0} & \binom{a_2+b_2-a_0-b_0}{a_2-a_0} & \cdots & \binom{a_{n-1}+b_{n-1}-a_0-b_0}{a_{n-1}-a_0} & \binom{a_n+b_n-a_0-b_0}{a_n-a_0} \\ 1 & \binom{a_2+b_2-a_1-b_1}{a_2-a_1} & \cdots & \binom{a_{n-1}+b_{n-1}-a_1-b_1}{a_{n-1}-a_1} & \binom{a_n+b_n-a_1-b_1}{a_n-a_1} \\ 0 & 1 & \cdots & \binom{a_{n-1}+b_{n-1}-a_1-b_2}{a_{n-1}-a_2} & \binom{a_n+b_n-a_2-b_2}{a_n-a_2} \\ \vdots & \vdots & \ddots & \vdots & \vdots \\ 0 & 0 & \cdots & \binom{a_{n-1}+b_{n-1}-a_{n-2}-b_{n-2}}{a_{n-1}-a_{n-2}} & \binom{a_n+b_n-a_{n-2}-b_{n-2}}{a_n-a_{n-2}} \\ 0 & 0 & \cdots & 1 & \binom{a_n+b_n-a_{n-1}-b_{n-1}}{a_n-a_{n-1}} \end{pmatrix}.$$

(b) Use the previous exercise to prove the following determinantal formula for the Catalan numbers:

$$C_n = \frac{(-1)^n}{2} \det \begin{pmatrix} \binom{2}{1} & \binom{4}{2} & \binom{6}{3} & \cdots & \binom{2n}{n} & \binom{2n+2}{n+1} \\ 1 & \binom{2}{1} & \binom{4}{2} & \cdots & \binom{2n-2}{n-1} & \binom{2n}{n} \\ 0 & 1 & \binom{2}{1} & \cdots & \binom{2n-4}{n-2} & \binom{2n-2}{n-1} \\ \vdots & \vdots & \vdots & \ddots & \vdots & \vdots \\ 0 & 0 & 0 & \cdots & \binom{2}{1} & \binom{4}{2} \\ 0 & 0 & 0 & \cdots & 1 & \binom{2}{1} \end{pmatrix}.$$

Chapter 4
The Ring of Symmetric Functions

4.1 The Cauchy Product

The expression $\prod_{i,j}(1 - x_i y_j)^{-1}$ is called *the Cauchy product*. We will compute it in two different ways and then use this computation to establish relations between various bases in the ring Λ_n.

Proposition 4.1 *The following formula holds:*

$$\prod_{i,j=1}^{n} \frac{1}{1 - x_i y_j} = \sum_{\lambda} h_\lambda(\mathbf{x}) m_\lambda(\mathbf{y}).$$

Here the sum is taken over all partitions λ of length $\leq n$.

Proof Recall that the generating function for h_k looks as follows:

$$H_{\mathbf{x}}(t) = \sum_{t=0}^{n} h_k(\mathbf{x})t^k = \prod_{i=1}^{\infty} \frac{1}{1 - x_i t}.$$

This implies that

$$\prod_{i,j=1}^{n} \frac{1}{1 - x_i y_j} = \prod_{j=1}^{n} H_{\mathbf{x}}(y_j)$$

$$= \prod_{j=1}^{n} \sum_{k=0}^{\infty} h_k(\mathbf{x}) y_j^k$$

$$= \sum_{k_1=0}^{\infty} \cdots \sum_{k_n=0}^{\infty} h_{k_1}(\mathbf{x}) \ldots h_{k_n}(\mathbf{x}) y_1^{k_1} \ldots y_n^{k_n} = (*)$$

Each multidegree (sequence of numbers) $k_1, k_2, \ldots, k_n$ can be obtained from a certain partition $\lambda = (\lambda_1, \ldots, \lambda_n)$ by rearranging its elements. Now put together the summands corresponding to a given λ and take their sum over all partitions:

E. Smirnov, A. Tutubalina, *Symmetric Functions: A Beginner's Course*,
Moscow Lectures 10, https://doi.org/10.1007/978-3-031-50341-2_4

$$(*) = \sum_{\lambda} h_{\lambda_1}(\mathbf{x}) \cdots h_{\lambda_n}(\mathbf{x}) \sum_{\sigma \in S(\lambda)} y_1^{\lambda_{\sigma(1)}} \cdots y_n^{\lambda_{\sigma(n)}} = \sum_{\lambda} h_{\lambda}(\mathbf{x}) m_{\lambda}(\mathbf{y}),$$

This proves the desired identity.   □

Theorem 4.2 (The first Cauchy formula) *The following formula holds:*

$$\prod_{i,j=1}^{n} \frac{1}{1 - x_i y_j} = \sum_{\lambda} s_{\lambda}(\mathbf{x}) s_{\lambda}(\mathbf{y}).$$

The sum is taken over all partitions λ with at most n parts.

Proof Let us compute the *Cauchy determinant* in two different ways.

$$\mathcal{A} = \det\left(\frac{1}{1 - x_i y_j}\right)_{i,j=1}^{n}.$$

The change of variables $r_i = -x_i, t_j = y_j^{-1}$ makes it homogeneous:

$$\mathcal{A} = \det\left(\frac{t_j}{t_j + r_i}\right)_{i,j=1}^{n} = \left(\prod_{j=1}^{n} t_j\right) \det\left(\frac{1}{t_j + r_i}\right)_{i,j=1}^{n}.$$

We have already computed this determinant (cf. Problem 2.2), but for the sake of completeness let us do it again. Multiplying the determinant by $\prod_{i,j=1}^{n}(t_j + r_i)$, we obtain a homogeneous polynomial of degree $n(n-1)$ (since the degree of the product is n^2, while the degree of the determinant equals $-n$). It is antisymmetric with respect to r_i and t_i, so it is divisible by $\prod_{i<j}(r_i - r_j) \prod_{i<j}(t_i - t_j)$. This is also a polynomial of degree $n(n-1)$, so their ratio is a constant. Comparing the coefficients in front of a certain monomial, we obtain that this ratio is equal to 1.

Summarizing, we have

$$\mathcal{A} = \left(\prod_{j=1}^{n} t_j\right)\left(\prod_{i,j=1}^{n} \frac{1}{t_j + r_i}\right)\left(\prod_{i<j}(r_i - r_j)(t_i - t_j)\right)$$

$$= \left(\prod_{j=1}^{n} \frac{1}{y_j}\right)\left(\prod_{i,j=1}^{n} \frac{y_j}{1 - x_i y_j}\right)\left(\prod_{i<j}(-x_i + x_j)\left(\frac{1}{y_i} - \frac{1}{y_j}\right)\right)$$

$$= \left(\prod_{j=1}^{n} \frac{1}{y_j^n}\right)\left(\prod_{i,j=1}^{n} \frac{y_j}{1 - x_i y_j}\right)\left(\prod_{i<j}(x_i - x_j)(y_i - y_j)\right)$$

$$= \frac{a_{\delta}(\mathbf{x}) a_{\delta}(\mathbf{y})}{\prod_{i,j=1}^{n}(1 - x_i y_j)}.$$

This determinant can be computed in a different way.

$$\det\left(\frac{1}{1-x_iy_j}\right)^n_{i,j=1} = \det\left(1+x_iy_j+x_i^2y_j^2+\cdots\right)^n_{i,j=1}$$

$$= \sum_{(a_1,\ldots,a_n)\in\mathbb{Z}^n_{\geq 0}}\det\left(x_i^{a_i}y_j^{a_i}\right)^n_{i,j=1}$$

$$= \sum_{(a_1,\ldots,a_n)\in\mathbb{Z}^n_{\geq 0}}x_1^{a_1}\ldots x_n^{a_n}\det\left(y_j^{a_i}\right)^n_{i,j=1}$$

$$= \sum_{(a_1,\ldots,a_n)\text{ pairwise distinct}}x_1^{a_1}\ldots x_n^{a_n}\det\left(y_j^{a_i}\right)^n_{i,j=1} = (*).$$

(We pass to the sum over n-tuples with pairwise distinct entries, because if two indices a_i are equal, the determinant $\det\left(y_j^{a_i}\right) = 0$ vanishes.) Rearrange this into a sum over *decreasing* $a_1,\ldots,a_n$, applying appropriate permutations:

$$(*) = \sum_{a_1>\cdots>a_n}\sum_{w\in S_n}x_1^{a_{w(1)}}\ldots x_n^{a_{w(n)}}(-1)^w\det\left(y_j^{a_i}\right)^n_{i,j=1}$$

$$= \sum_{a_1>\cdots>a_n}\det\left(x_i^{a_j}\right)^n_{i,j=1}\det\left(y_j^{a_i}\right)^n_{i,j=1}.$$

Finally, observe that $a = (a_1 > \cdots > a_n)$ can be presented as a sum $\lambda + \delta$, where λ is a partition. So we have

$$\det\left(\frac{1}{1-x_iy_j}\right)^n_{i,j=1} = \sum_\lambda a_{\lambda+\delta}(\mathbf{x})a_{\lambda+\delta}(\mathbf{y}) = a_\delta(\mathbf{x})a_\delta(\mathbf{y})\sum_\lambda s_\lambda(\mathbf{x})s_\lambda(\mathbf{y}),$$

which implies the first Cauchy formula. $\qquad\qquad\qquad\qquad\qquad\qquad\qquad\square$

4.2 Ring of Symmetric Functions as the Projective Limit of Λ_n

Consider the space $\Lambda_n^{(k)}$ of homogeneous symmetric polynomials of degree k in n variables. Assume that $n > k$. Then the polynomials $m_\lambda(x_1,\ldots,x_n)$ with $|\lambda| = k$ form a basis of this space. This means that the dimension of $\Lambda_n^{(k)}$ is equal to the number $p(k)$ of partitions of k.

Note that this dimension is independent of n. Moreover, almost all algebraic identities for symmetric polynomials we have seen so far (the determinantal identities for e_k, h_k, and p_k, the Jacobi–Trudi identities, the Pieri formulas) are also independent of the number of variables.

This gives us the following idea: we can consider symmetric "polynomials" (in fact, they will be formal power series) in infinitely many variables. For example,

$$p_k = x_1^k + x_2^k + \cdots \text{ or } e_2 = x_1x_2 + x_1x_3 + x_2x_3 + x_1x_4 + \cdots$$

We can truncate such a series at any n by setting $x_{n+1} = x_{n+2} = \cdots = 0$ and get a familiar object: a symmetric polynomial in n variables.

Let us formalize this idea. For this we shall need one notion from category theory.

Definition 4.3 Take a family of *objects* M_n from a certain category, indexed by the natural numbers, and *morphisms* $f_n \colon M_{n+1} \to M_n$,

$$\cdots \xrightarrow{f_{n+1}} M_{n+1} \xrightarrow{f_n} M_n \xrightarrow{f_{n-1}} \cdots \xrightarrow{f_2} M_2 \xrightarrow{f_1} M_1.$$

Then there exists their *projective limit* $M = \varprojlim M_n$ with a family of *projections* $\pi_n \colon M \to M_n$,

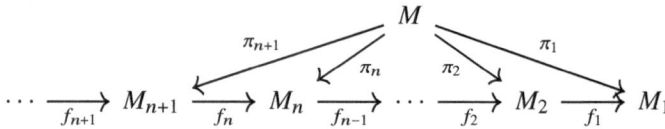

such that this diagram is commutative: $f_n \circ \pi_{n+1} = \pi_n$, and the projective limit M satisfies the *universal property*. This means that for any object X and a collection of morphisms $\xi_n \colon X \to M_n$, such that $f_n \circ \xi_{n+1} = \xi_n$, these morphisms "factor through M": there exists a unique map $\alpha \colon X \to M$ such that $\pi_n \circ \alpha = \xi_n$.

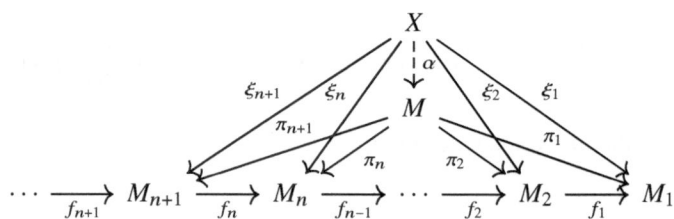

Remark 4.4 The universality property guarantees that the projective limit is unique up to an isomorphism.

Now we can apply this definition to symmetric polynomials. The graded rings of symmetric polynomials in n variables form a decreasing sequence

$$\cdots \xrightarrow{\zeta_{n+1}} \Lambda_{n+1} \xrightarrow{\zeta_n} \Lambda_n \xrightarrow{\zeta_{n-1}} \cdots \xrightarrow{\zeta_2} \Lambda_2 \xrightarrow{\zeta_1} \Lambda_1,$$

where

$$\zeta_{n+1} \colon f(x_1, x_2, \ldots, x_n, x_{n+1}) \mapsto f(x_1, x_2, \ldots, x_n, 0).$$

Hence we can define their projective limit.

Definition 4.5 The *ring of symmetric functions* Λ is defined as the projective limit of Λ_n:

$$\Lambda := \varprojlim \Lambda_n.$$

The ring Λ can be described explicitly, without using universal objects.

Definition 4.6 A *homogeneous symmetric function* of degree k is a formal power series

$$f(\mathbf{x}) = \sum_{\substack{\alpha=(\alpha_1,\alpha_2,\dots) \\ \alpha_i \in \mathbb{Z}_{\geq 0} \\ \alpha_1+\alpha_2+\cdots=k}} c_\alpha \mathbf{x}^\alpha$$

in countably many variables $\mathbf{x} = (x_1, x_2, \dots)$ with coefficients $c_\alpha \in \mathbb{Z}$ (sometimes we will use $c_\alpha \in \mathbb{Q}$), such that for any sequences α and α' obtained from each other by a permutation, the coefficients c_α and $c_{\alpha'}$ coincide.

Denote the set of homogeneous symmetric functions of degree k by $\Lambda^{(k)}$, assuming $\Lambda^{(0)} = \mathbb{Z}$.

It is clear that for any $f, g \in \Lambda^{(k)}, a, b \in \mathbb{Z}$ the function $af + bg$ also belongs to $\Lambda^{(k)}$. So $\Lambda^{(k)}$ is a $\mathbb{Z}$-module.

It is also easy to check that for any $f \in \Lambda^{(k)}$ and $g \in \Lambda^{(m)}$ we have a well-defined product $fg \in \Lambda^{(k+m)}$. So the direct sum of modules $\Lambda^{(k)}$ forms a graded ring Λ:

$$\Lambda := \bigoplus_{k=0}^{\infty} \Lambda^{(k)}.$$

This is an explicit construction of the ring of symmetric functions.

All five standard bases in the rings Λ_n agree with the maps ζ_n:

$$\zeta_n : s_\lambda(x_1, \dots, x_n, x_{n+1}) \mapsto s_\lambda(x_1, \dots, x_n, 0) = s_\lambda(x_1, \dots, x_n),$$

the same holds for $m_\lambda, e_\lambda, h_\lambda, p_\lambda$. Hence we can define monomial, elementary, complete symmetric functions, Newton power sums and Schur functions as projective limits of the respective bases in Λ_n. Obviously, all these families will be bases in Λ.

Let us give explicit forms of these functions. Let $\lambda = (\lambda_1, \lambda_2, \dots)$ be a partition, with $\lambda_N = 0$ for $N \gg 0$. Then

$$m_\lambda(\mathbf{x}) = \sum_\alpha x_1^{\alpha_1} x_2^{\alpha_2} \dots,$$

where $\alpha = (\alpha_1, \alpha_2, \dots)$ is obtained from $(\lambda_1, \lambda_2, \dots)$ by a certain permutation. It is more convenient to define the functions $e_\lambda, h_\lambda, p_\lambda$ and s_λ for a partition $\lambda = (\lambda_1, \dots, \lambda_m)$ without zeros:

$$e_k(\mathbf{x}) = \sum_{1 \leq i_1 < i_2 < \cdots < i_k} x_{i_1} x_{i_2} \dots x_{i_k}, \qquad e_\lambda = e_{\lambda_1} \dots e_{\lambda_m};$$

$$h_k(\mathbf{x}) = \sum_{1 \leq i_1 \leq i_2 \leq \cdots \leq i_k} x_{i_1} x_{i_2} \dots x_{i_k}, \qquad h_\lambda = h_{\lambda_1} \dots h_{\lambda_m};$$

$$p_k(\mathbf{x}) = \sum_{i=1}^{\infty} x_i^k, \qquad p_\lambda = p_{\lambda_1} \dots p_{\lambda_m};$$

$$s_\lambda(\mathbf{x}) = \sum_{T \in \mathrm{SSYT}_\lambda} \mathbf{x}^T,$$

where T ranges over the set of all semistandard Young tableaux of shape λ filled by arbitrary positive integers.

4.3 The Involution ω

In the first chapter we have seen that e_k are expressed in terms of h_ℓ with the same coefficients which appear in the expression of h_k in terms of e_ℓ. This implies that the automorphism

$$\omega: \Lambda \to \Lambda, \qquad \omega(h_\lambda) = e_\lambda$$

is an involution: $\omega^2 = \mathrm{id}$.

Comparing the first (Theorem 2.11) and the second (Problem 3.5) Jacobi–Trudi identities, we see that $\omega(s_\lambda) = s_{\lambda'}$.

It turns out (Problem 4.2) that p_λ are eigenvectors of this automorphism: $\omega(p_\lambda) = \varepsilon_\lambda p_\lambda$, where ε_λ is the sign of a permutation with cyclic type λ.

Definition 4.7 *Forgotten symmetric functions are defined as* $f_\lambda = \omega(m_\lambda)$. *They also form a basis in the ring* Λ.

Exercise 4.8 Compute $f_{(2,1)}$.

4.4 The Hall Inner Product

Definition 4.9 Consider a bilinear form

$$\langle \cdot, \cdot \rangle: \Lambda \times \Lambda \to \mathbb{R}$$

in the space of symmetric functions Λ defined as follows:

$$\langle m_\lambda, h_\mu \rangle = \delta_{\lambda\mu}$$

(here $\delta_{\lambda\mu}$ stands for the Kronecker symbol). In other words, the bases h_λ and m_λ are dual with respect to this form.

In the beginning of this chapter we have seen that the bases m_λ and h_λ are related by means of the Cauchy product:

$$\prod_{i,j=1}^{\infty} \frac{1}{1 - x_i y_j} = \sum_\lambda h_\lambda(\mathbf{x}) m_\lambda(\mathbf{y}).$$

Here we take the limit as n tends to infinity and pass to the sum over all partitions λ.

It turns out that the same formula holds for any pair of dual bases.

Lemma 4.10 *Let* $\{u_\lambda\}$ *and* $\{v_\lambda\}$ *be bases in the ring of symmetric functions* Λ, *indexed by partitions* λ, *such that for* $\lambda \vdash n$ *the functions* u_λ, v_λ *belong to* $\Lambda^{(n)}$. *Then the following are equivalent:*

- $\langle u_\lambda, v_\mu \rangle = \delta_{\lambda\mu}$ *for any partitions* λ, μ.
- $\sum_\lambda u_\lambda(\mathbf{x}) v_\lambda(\mathbf{y}) = \prod_{i,j} \frac{1}{1-x_i y_j}$.

Proof Express m_λ and h_μ in terms of u_ρ and v_ν, respectively:

$$m_\lambda = \sum_\rho \xi_{\lambda\rho} u_\rho;$$

$$h_\mu = \sum_\nu \eta_{\mu\nu} v_\nu.$$

Then we have

$$\delta_{\lambda\mu} = \langle m_\lambda, h_\mu \rangle = \sum_{\rho,\nu} \xi_{\lambda\rho} \langle u_\rho, v_\nu \rangle \eta_{\mu\nu}. \tag{4.1}$$

Introduce matrices ξ, η, and A with rows and columns indexed by partitions: $\xi = (\xi_{\lambda\rho})_{\lambda,\rho}$, $\eta = (\eta_{\mu\nu})_{\mu,\nu}$ and $A = (\langle u_\rho, v_\nu \rangle)_{\rho,\nu}$. The equation (4.1) is equivalent to the condition $\xi A \eta^T = E$.

Next, the bases $\{u_\lambda\}$ and $\{v_\lambda\}$ are dual, which is equivalent to the following:

$$A = E \iff \xi \eta^T = E$$
$$\iff \xi^T \eta = E$$
$$\iff \sum_\lambda \xi_{\lambda\rho} \eta_{\lambda\nu} = \delta_{\rho\nu}.$$

It remains to observe that

$$\prod_{i,j} \frac{1}{1 - x_i y_j} = \sum_\lambda m_\lambda(\mathbf{x}) h_\lambda(\mathbf{y})$$

$$= \sum_\lambda \left(\sum_\rho \xi_{\lambda\rho} u_\rho(\mathbf{x}) \right) \left(\sum_\nu \eta_{\lambda\nu} v_\nu(\mathbf{y}) \right)$$

$$= \sum_{\rho,\mu} \left(\sum_\lambda \xi_{\lambda\rho} \eta_{\lambda\nu} \right) u_\rho(\mathbf{x}) v_\nu(\mathbf{y}),$$

and the functions $u_\rho(\mathbf{x}) v_\nu(\mathbf{y})$ are linearly independent. So $\sum_\lambda u_\lambda(\mathbf{x}) v_\lambda(\mathbf{y}) = \prod_{i,j} \frac{1}{1-x_i y_j}$ if and only if $\sum_\lambda \xi_{\lambda\rho} \eta_{\lambda\nu} = \delta_{\rho\nu}$, i.e., the bases $\{u_\lambda\}$ and $\{v_\lambda\}$ are dual. $\qquad\square$

Since the Cauchy formula holds, we have $\langle s_\lambda, s_\mu \rangle = \delta_{\lambda\mu}$. The involution ω preserves the scalar product $\langle s_\lambda, s_\mu \rangle$. Summarizing, we obtain the following corollary.

Corollary 4.11 (1) *The Schur functions form an orthonormal basis in Λ, and the inner product $\langle \cdot, \cdot \rangle$ is positive definite.*
(2) *The involution ω is an isometry of the space Λ, i.e. $\langle \omega f, \omega g \rangle = \langle f, g \rangle$.*

This inner product is usually called the *Hall inner product*.

Exercise 4.12 (cf. Problems 1.4 and 1.5) Recall the combinatorial expression of coefficients $M_{\lambda\mu}$ and $N_{\lambda\mu}$ in the equations

$$e_\lambda = \sum_\mu M_{\lambda\mu} m_\mu \qquad h_\lambda = \sum_\mu N_{\lambda\mu} m_\mu.$$

(a) Using these coefficients, express the inner products $\langle h_\lambda, h_\mu \rangle$, $\langle e_\lambda, e_\mu \rangle$, $\langle e_\lambda, h_\mu \rangle$.
(b) Compute $\langle e_{(n)}, h_{(n)} \rangle$, $\langle e_{(1^n)}, h_{(n)} \rangle$, and $\langle e_{(1^n)}, e_{(1^n)} \rangle$ explicitly.

4.5 One More Proof of Littlewood's Theorem

The Cauchy identity provides an algebraic proof of Littlewood's theorem (Theorem 3.4).

Using the Pieri formulas, we can see that the coefficients appearing in the expression of h_λ in the basis of Schur polynomials s_μ are the Kostka numbers $K_{\mu\lambda}$ (cf. Problem 3.4):

$$h_\lambda = \sum_\mu K_{\mu\lambda} s_\lambda.$$

Substituting this expression into the Cauchy formula:

$$\sum_\lambda h_\lambda(\mathbf{x}) m_\lambda(\mathbf{y}) = \sum_\lambda s_\lambda(\mathbf{x}) s_\lambda(\mathbf{y}),$$

we obtain the following identity:

$$\sum_{\lambda,\mu} K_{\mu\lambda} s_\mu(\mathbf{x}) m_\lambda(\mathbf{y}) = \sum_\lambda s_\lambda(\mathbf{x}) s_\lambda(\mathbf{y}).$$

Now change the summation indices:

$$\sum_{\lambda,\mu} K_{\lambda\mu} s_\lambda(\mathbf{x}) m_\mu(\mathbf{y}) = \sum_\lambda s_\lambda(\mathbf{x}) s_\lambda(\mathbf{y})$$

and take the inner product of both sides with $s_\lambda(\mathbf{x})$ (or just use the linear independence of $s_\lambda(\mathbf{x})$). We obtain the identity

$$s_\lambda(\mathbf{y}) = \sum_\mu K_{\lambda\mu} m_\mu(\mathbf{y}),$$

which is exactly the statement of Littlewood's theorem.

4.6 Relations Between Bases of Λ

Let us summarize what we know so far about different bases in the ring of symmetric functions Λ. The relations between them are shown in the diagram below.

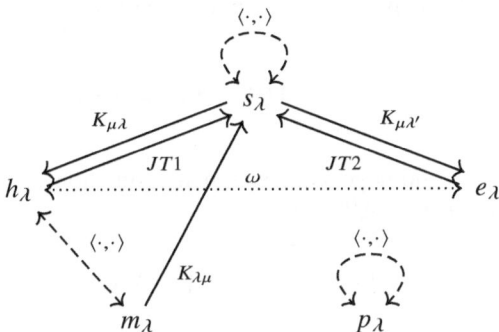

The Schur functions s_λ are expressed in terms of h_λ and e_λ by means the first (Theorem 2.11) and the second (Theorem 3.5) Jacobi–Trudi identities. The Pieri formulas (2.14) provide expressions for h_λ and e_λ in terms of Schur functions and Kostka numbers.

Littlewood's theorem claims that the Schur functions s_λ can be expressed as a linear combination of monomial symmetric functions m_μ with coefficients $K_{\lambda\mu}$.

The involution $\omega\colon \Lambda \to \Lambda$ interchanges e_λ and h_λ. The Hall inner product $\langle \cdot, \cdot \rangle$ on Λ is defined by the condition that the bases h_λ and m_λ are dual. With respect to this product the basis s_λ is orthonormal, and the basis p_λ is orthogonal (but not orthonormal, cf. Problem 4.4 (b)).

4.7 Problems

4.1 Prove the identities

(a) $\prod_{i,j}(1 + x_i y_j) = \sum_\lambda e_\lambda(\mathbf{x}) m_\lambda(\mathbf{y})$;

(b) $\prod_{i,j}(1 + x_i y_j) = \sum_\lambda s_{\lambda'}(\mathbf{x}) s_\lambda(\mathbf{y})$ (the second Cauchy formula).

4.2 Let $\lambda = (1^{m_1}, 2^{m_2}, 3^{m_3}, \dots)$ be a partition. Define

$$z_\lambda = 1^{m_1} \cdot m_1! \cdot 2^{m_2} \cdot m_2! \cdot 3^{m_3} \cdot m_3! \cdot \dots.$$

Prove the equalities

(a) $H(t) = \sum_\lambda z_\lambda^{-1} p_\lambda t^{|\lambda|}$;

(b) $E(t) = \sum_\lambda \varepsilon_\lambda z_\lambda^{-1} p_\lambda t^{|\lambda|}$, where $\varepsilon_\lambda = (-1)^{m_2+m_4+\cdots} = (-1)^{|\lambda|-\ell(\lambda)}$ is the sign of a permutation with cycle type λ.

4.3 Prove that

$$\prod_{i,j} \frac{1}{1 - x_i y_j} = \exp \sum_{n \geq 1} \frac{1}{n} p_n(\mathbf{x}) p_n(\mathbf{y}) = \sum_\lambda z_\lambda^{-1} p_\lambda(\mathbf{x}) p_\lambda(\mathbf{y});$$

$$\prod_{i,j} (1 + x_i y_j) = \exp \sum_{n \geq 1} \frac{(-1)^{n-1}}{n} p_n(\mathbf{x}) p_n(\mathbf{y}) = \sum_\lambda \varepsilon_\lambda z_\lambda^{-1} p_\lambda(\mathbf{x}) p_\lambda(\mathbf{y}).$$

4.4 (a) Using the previous exercise, show that $\omega(p_\lambda) = \varepsilon_\lambda p_\lambda$.

(b) Show that $\langle p_\lambda, p_\mu \rangle = z_\lambda \delta_{\lambda\mu}$.

4.5 Let $f \in \Lambda^{(n)}$ be a homogeneous symmetric function of degree n. Define a symmetric function $f_k \in \Lambda^{(nk)}$ as

$$f_k(x_1, x_2, \dots) = f\left(x_1^k, x_2^k, \dots\right).$$

Prove that

$$\omega f_k = (-1)^{n(k-1)} (\omega f)_k.$$

4.6 (a) Let

$$C(\mathbf{x}, \mathbf{y}) = \prod_{i,j} \frac{1}{1 - x_i y_j}.$$

Show that for any symmetric function $f \in \Lambda$ we have

$$\langle C(\mathbf{x}, \mathbf{y}), f(\mathbf{x}) \rangle = f(\mathbf{y})$$

(the inner product is taken with respect to the $\mathbf{x}$-variables).

(b) Find a function $D(\mathbf{x}, \mathbf{y})$ such that

$$\langle D(\mathbf{x}, \mathbf{y}), f(\mathbf{x}) \rangle = (\omega f)(\mathbf{y})$$

for any function $f \in \Lambda$.

Chapter 5
The Number of Young Tableaux

As we have seen before, a Schur polynomial is equal to the sum of monomials for Young tableaux of a given shape. In this chapter we address the following question: what is the number of such tableaux?

Denote the number of semistandard Young tableaux of shape λ, filled by numbers not exceeding n, by $K_\lambda(n)$,

$$K_\lambda(n) = s_\lambda(\underbrace{1, 1, \ldots, 1}_{n \text{ 1s}}) = |\text{SSYT}_\lambda(n)|.$$

We will need one more definition.

Definition 5.1 A semistandard Young tableau of shape λ is called a *standard Young tableau* if all the entries from 1 to $|\lambda|$ appear in it exactly once. In other words, a standard tableau is a semistandard tableau of weight $\mu = (1, 1, \ldots, 1)$.

1	3	4	7
2	5	9	
6	8		

Fig. 5.1: A standard tableau of shape $\lambda = (4, 3, 2)$

We will denote the set of standard Young tableaux of a given shape by SYT_λ.

Let $K_\lambda = |\text{SYT}_\lambda| = K_{\lambda,(1^{|\lambda|})}$ be the number of standard tableaux of shape λ.

5.1 The Number of Young Tableaux for Large n

Let us evaluate the growth of $K_\lambda(n)$ for large n. For this we will use the Cauchy formula

$$\prod_{i,j=1}^{n} \frac{1}{1 - x_i y_j} = \sum_\lambda s_\lambda(\mathbf{x}) s_\lambda(\mathbf{y}).$$

© The Author(s), under exclusive license to Springer Nature Switzerland AG 2024
E. Smirnov, A. Tutubalina, *Symmetric Functions: A Beginner's Course*,
Moscow Lectures 10, https://doi.org/10.1007/978-3-031-50341-2_5

Evaluate it at $y_1 = \cdots = y_n = \frac{1}{n}$. Since Schur polynomials are homogeneous of degree $|\lambda|$, we have $s_\lambda\left(\frac{1}{n}, \ldots, \frac{1}{n}\right) = K_\lambda(n)n^{-|\lambda|}$. We obtain the following equality:

$$\prod_{i=1}^{n}\left(1 - \frac{x_i}{n}\right)^{-n} = \sum_\lambda s_\lambda(\mathbf{x})K_\lambda(n)n^{-|\lambda|}.$$

Now take the limit as n tends to infinity:

$$\lim_{n\to\infty}\sum_\lambda s_\lambda(\mathbf{x})K_\lambda(n)n^{-|\lambda|} = \lim_{n\to\infty}\prod_{i=1}^{n}\left(1 - \frac{x_i}{n}\right)^{-n}$$
$$= \exp(x_1 + x_2 + \cdots) = \exp(h_1)$$
$$= \sum_{k=0}^{\infty}\frac{h_1^k}{k!} = \sum_{k=0}^{\infty}\frac{h_{(1^k)}}{k!}.$$

The functions h_μ are expressed in the basis of Schur functions with coefficients equal to Kostka numbers (cf. Problem 3.4):

$$h_\mu = \sum_\lambda K_{\lambda\mu}s_\lambda.$$

Applying this formula to $h_{(1^k)}$, we see that

$$h_{(1^k)} = \sum_{\lambda\vdash k}K_{\lambda,(1^k)}s_\lambda = \sum_{\lambda\vdash k}K_\lambda s_\lambda.$$

So we obtain that

$$\sum_\lambda\left(\lim_{n\to\infty}K_\lambda(n)n^{-|\lambda|}\right)s_\lambda = \sum_\lambda\frac{K_\lambda}{|\lambda|!}s_\lambda.$$

Using the linear independence of s_λ, we obtain the following corollary from the Cauchy identity.

Corollary 5.2 *The following identity holds:*

$$\lim_{n\to\infty}\frac{K_\lambda(n)}{n^{|\lambda|}} = \frac{K_\lambda}{|\lambda|!}.$$

This formula can be rewritten in a slightly different way. The number of Young tableaux λ filled by *pairwise distinct* numbers not exceeding n is equal to $\binom{n}{|\lambda|}K_\lambda$.

$$\lim_{n\to\infty}\frac{K_\lambda(n)}{\binom{n}{|\lambda|}K_\lambda} = \lim_{n\to\infty}\frac{n^{|\lambda|}}{\binom{n}{|\lambda|}|\lambda|!} = \lim_{n\to\infty}\frac{n^{|\lambda|}}{n(n-1)\cdots(n-|\lambda|+1)} = 1.$$

Informally, for n large, "almost all" semistandard Young tableaux are filled by pairwise distinct numbers.

5.2 q-Binomial Coefficients

Before we proceed with Young tableaux, we will make a short detour and discuss the so-called q-binomial, or Gaussian binomial, coefficients.

Definition 5.3 The *q-binomial coefficient* $\begin{bmatrix} n \\ k \end{bmatrix}$ is defined as the generating function for the weights of Young diagrams that fit into the rectangle of size $k \times (n - k)$:

$$\begin{bmatrix} n \\ k \end{bmatrix} = \sum_{\lambda \subset k \times (n-k)} q^{|\lambda|}.$$

Some properties of q-binomial coefficients immediately follow from the definition.

- $\begin{bmatrix} n \\ k \end{bmatrix}_{q=1} = \binom{n}{k}$ (the number of Young diagrams inside a rectangle is equal to the number of paths from the southeast corner to the northwest corner);
- $\begin{bmatrix} m+n \\ n \end{bmatrix} = \begin{bmatrix} m+n \\ m \end{bmatrix}$;
- the polynomial $P(q) = \begin{bmatrix} n \\ k \end{bmatrix}$ is palindromic: $P(q^{-1}) = q^{-k(n-k)} P(q)$ (prove this).

Ordinary binomial coefficients satisfy the recurrence relation $\binom{n+1}{k} = \binom{n}{k} + \binom{n}{k-1}$. This relation has a q-analog.

Proposition 5.4 *q-binomial coefficients satisfy the relation*

$$\begin{bmatrix} n + 1 \\ k \end{bmatrix} = \begin{bmatrix} n \\ k \end{bmatrix} + q^{n-k+1} \begin{bmatrix} n \\ k - 1 \end{bmatrix}.$$

Proof The left-hand side indexes all Young diagrams inside the rectangle of size $k \times (n - k + 1)$. We distinguish between two cases:

- Young diagrams with less than $n - k + 1$ rows. All such diagrams fit into the rectangle of size $k \times (n - k)$, and the generating function for their areas is equal to $\begin{bmatrix} n \\ k \end{bmatrix}$.
- Young diagrams with exactly $n-k+1$ rows. If we remove its first column (consisting of exactly $n - k + 1$ boxes), the remaining diagram fits into the rectangle of size $(k - 1) \times (n - k + 1)$. Hence the generating function for the areas of such diagrams is equal to $q^{n-k+1} \begin{bmatrix} n \\ k-1 \end{bmatrix}$.   □

Ordinary binomial coefficients can also be computed using an explicit formula: $\binom{n}{k} = \frac{n!}{k!(n-k)!}$. It turns out that its analog also holds for q-binomial coefficients. We shall need q-analogues of natural numbers and factorials.

Definition 5.5 (1) Polynomials $[n] = 1 + q + q^2 + \cdots + q^{n-1} = \frac{1-q^n}{1-q}$ are called *q-numbers*.
(2) *q-factorial* $[n]!$ is the product of several consecutive q-numbers, starting from 1, that is $[1] \cdot [2] \cdots \cdots [n] = [n]!$. We also formally set $[0]! = 1$.
Clearly, the q-numbers and q-factorials for $q = 1$ are the ordinary natural numbers and factorials.

Proposition 5.6 *The q-binomial coefficients satisfy the relation*

$$\begin{bmatrix} n \\ k \end{bmatrix} = \frac{[n]!}{[k]![n-k]!}.$$

Exercise 5.7 Prove this, using the recurrence relation from Proposition 5.4.

Exercise 5.8 Compute $\begin{bmatrix} 4 \\ 2 \end{bmatrix}$ and $\begin{bmatrix} 5 \\ 2 \end{bmatrix}$ in two ways, combinatorially and algebraically, and check that the result is the same.

There is a direct relation between q-binomial coefficients and symmetric polynomials e_k and h_k, respectively.

Proposition 5.9 *q-binomial coefficients are* principal specializations *of complete symmetric polynomials:*

$$\begin{bmatrix} n+k-1 \\ k \end{bmatrix} = h_k\left(1, q, q^2, \ldots, q^{n-1}\right).$$

Proof According to Proposition 3.8, the complete symmetric polynomial $h_k(x_1, \ldots, x_n)$ equals the sum of monomials that correspond to paths inside a $k \times (n-1)$ rectangle, with the weight x_i at horizontal segments in the i-th row, counted from below.

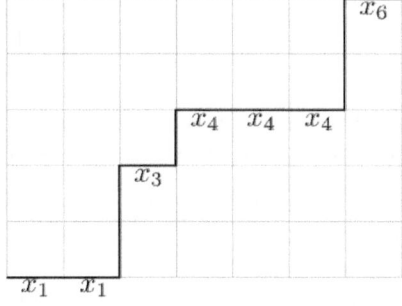

Every horizontal segment with the weight x_i has $i-1$ boxes underneath. So if a path corresponds to the monomial $x_{i_1} x_{i_2} \ldots x_{i_k}$, the area of the Young diagram under this path is equal to $(i_1 - 1) + (i_2 - 1) + \cdots + (i_k - 1)$.

This means that $h_k\left(1, q, q^2, \ldots, q^{n-1}\right)$ is the generating function for the area of Young diagrams, hence the proposition. (Formally, we are counting Young diagrams rotated by 180 degrees and starting from the southeast corner of the rectangle, but this obviously does not make any difference).   □

Exercise 5.10 Prove that

$$\begin{bmatrix} n \\ k \end{bmatrix} = q^{-k(k-1)/2} e_k\left(1, q, q^2, \ldots, q^{n-1}\right).$$

5.3 Principal Specializations of Schur Polynomials

Now let us compute the principal specialization of a Schur polynomial $s_\lambda\left(1, q, q^2, \ldots, q^{n-1}\right)$. For this we are going to use the definition of a Schur polynomial as the ratio of two determinants. This computation was first carried out by Dudley E. Littlewood and Archibald Read Richardson in [LR35].

Let $\lambda = (\lambda_1, \ldots, \lambda_n)$. Then

$$s_\lambda (x_1, \ldots, x_n) = \frac{\det \left(x_j^{\lambda_i + n - i}\right)}{\prod_{i<j} (x_i - x_j)}.$$

Now specialize at $x_i = q^{i-1}$:

$$s_\lambda \left(1, q, q^2, \ldots, q^{n-1}\right) = \frac{\det \left(q^{(j-1)(\lambda_i + n - i)}\right)}{\prod_{i<j} \left(q^{i-1} - q^{j-1}\right)}.$$

The numerator of this fraction is the Vandermonde determinant for variables $q^{\lambda_i + n - i}$. Here it is written upside-down, with the 1s in the last row instead of the first one. Thus

$$\begin{aligned} s_\lambda \left(1, q, q^2, \ldots, q^{n-1}\right) &= \frac{\prod_{i<j} \left(q^{\lambda_j + n - j} - q^{\lambda_i + n - i}\right)}{\prod_{i<j} \left(q^{i-1} - q^{j-1}\right)} \\ &= \frac{\prod_{i<j} q^{\lambda_j + n - j} \left(1 - q^{(\lambda_i + n - i) - (\lambda_j + n - j)}\right)}{\prod_{i<j} \left(q^{i-1} - q^{j-1}\right)}. \end{aligned} \qquad (5.1)$$

Let us compute the common power of N in front of q in the numerator. When we take the product over $i < j$, the factor q^j appears $j - 1$ times. Hence we have

$$N = \sum_{j=1}^{n} (\lambda_j + n - j)(j - 1) = n(\lambda) + \sum_{j=1}^{n} (n - j)(j - 1) = n(\lambda) + \frac{n(n-1)(n-2)}{6}.$$

Here $n(\lambda) = \sum_{j=1}^{n} \lambda_j (j - 1)$ is the sum of numbers obtained by filling the first row of the Young diagram with zeros, the second row with 1s, the third row with 2s, etc.

Now rewrite the denominator in (5.1) as follows:

$$\begin{aligned} \prod_{i<j} \left(q^{i-1} - q^{j-1}\right) &= \prod_{i<j} q^{i-1} \left(1 - q^{j-i}\right) \\ &= q^{\sum_{i=1}^{n}(i-1)(n-i)} \prod_{i<j} \left(1 - q^{j-i}\right) \\ &= q^{n(n-1)(n-2)/6} \prod_{i<j} \left(1 - q^{j-i}\right). \end{aligned}$$

So we have

$$s_\lambda\left(1, q, q^2, \ldots, q^{n-1}\right) = q^{n(\lambda)} \cdot \frac{\prod_{i<j}\left(1 - q^{(\lambda_i+n-i)-(\lambda_j+n-j)}\right)}{\prod_{i<j}\left(1 - q^{j-i}\right)}. \tag{5.2}$$

Further we will need one more definition and a lemma about Young diagrams.

Definition 5.11 Let $\lambda = (\lambda_1, \ldots, \lambda_n)$ be a partition, and let $\lambda' = \left(\lambda'_1, \ldots, \lambda'_m\right)$ be its conjugate. Then for a box of the Young diagram with the coordinates (i, j) (in the i-th row and the j-th column) the number

$$h(i, j) = \lambda_i - i + \lambda'_j - j + 1$$

is called its *hook length*. This number is indeed the length of the hook starting at the box (i, j), with an "arm" going right and the "leg" going down (cf. Fig. 5.2).

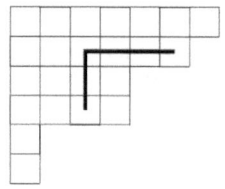

Fig. 5.2: For this partition $h(2, 3) = 6$

Lemma 5.12 *Let* $\lambda = (\lambda_1, \ldots, \lambda_n)$ *be a partition (possibly ending with zeros) and* $\lambda' = \left(\lambda'_1, \ldots, \lambda'_m\right)$ *be a conjugate partition (possibly also ending with zeros). Then*

$$\{m - 1 + i - \lambda_i \mid 1 \le i \le n\} \cup \left\{\lambda'_j + m - j \mid 1 \le j \le m\right\} = \{0, 1, 2, \ldots, m + n - 1\}. \tag{5.3}$$

Proof Consider an $m \times n$ rectangle and a Young diagram λ in it (cf. Fig. 5.3). Take the path from the northeast to the southwest corner going along the edge of this diagram. Let us index the steps of this path by numbers from 0 to $m + n - 1$. Then the horizontal steps correspond to the numbers $\lambda'_j + m - j$ (for $m \ge j \ge 1$), and the vertical ones correspond to the numbers $m - 1 + i - \lambda_i$ (for $1 \le i \le n$). This implies the lemma.   □

Now we return to the computation of the numerator in (5.2). Take $m = \lambda_1$, remove from the first set the zero corresponding to $i = 1$, and take the product of $(1 - q^t)$ for all t in the sets in (5.3):

$$\prod_{j=1}^{\lambda_1}\left(1 - q^{\lambda'_j+\lambda_1-j}\right) \cdot \prod_{j=2}^{n}\left(1 - q^{\lambda_1-1+j-\lambda_j}\right) = \prod_{j=1}^{\lambda_1+n-1}\left(1 - q^j\right).$$

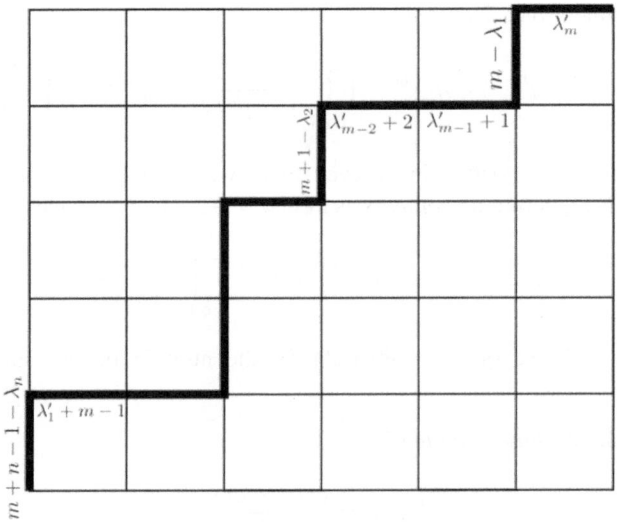

Fig. 5.3: A Young diagram λ in an $m \times n$ rectangle

Now rewrite the left-hand side in a slightly different way:

$$\prod_{j=1}^{\lambda_1} \left(1 - q^{h(1,j)}\right) \cdot \prod_{j=2}^{n} \left(1 - q^{(\lambda_1+n-1)-(\lambda_j+n-j)}\right) = \prod_{j=1}^{\lambda_1+n-1} \left(1 - q^j\right).$$

The same formulas can be written for the partitions $(\lambda_i, \lambda_{i+1}, \ldots, \lambda_n)$. Let us take the product of all of them:

$$\prod_{i=1}^{n} \left(\prod_{j=1}^{\lambda_i} \left(1 - q^{h(i,j)}\right) \cdot \prod_{j=i+1}^{n} \left(1 - q^{(\lambda_i+n-i)-(\lambda_j+n-j)}\right) \right) = \prod_{i=1}^{n} \prod_{j=1}^{\lambda_i+n-i} \left(1 - q^j\right),$$

$$\prod_{(i,j)\in\lambda} \left(1 - q^{h(i,j)}\right) \cdot \prod_{i<j} \left(1 - q^{(\lambda_i+n-i)-(\lambda_j+n-j)}\right)$$

$$= \prod_{i=1}^{n} \left(\prod_{j=1}^{n-i} \left(1 - q^j\right) \cdot \prod_{j=1}^{\lambda_i} \left(1 - q^{n-i+j}\right) \right),$$

$$\prod_{(i,j)\in\lambda} \left(1 - q^{h(i,j)}\right) \cdot \prod_{i<j} \left(1 - q^{(\lambda_i+n-i)-(\lambda_j+n-j)}\right)$$

$$= \prod_{i<j} \left(1 - q^{j-i}\right) \cdot \prod_{(i,j)\in\lambda} \left(1 - q^{n-i+j}\right).$$

Substituting this into (5.2), we get

$$s_\lambda \left(1, q, q^2, \ldots, q^{n-1}\right) = q^{n(\lambda)} \cdot \prod_{(i,j) \in \lambda} \frac{1 - q^{n-i+j}}{1 - q^{h(i,j)}} = q^{n(\lambda)} \prod_{(i,j) \in \lambda} \frac{[n - i + j]}{[h(i,j)]}.$$

This formula can be written in an even nicer way as follows. For a box x with coordinates (i, j), define its *contents* as $c(x) = j - i$. Then we have

$$s_\lambda \left(1, q, q^2, \ldots, q^{n-1}\right) = q^{n(\lambda)} \prod_{x \in \lambda} \frac{[n + c(x)]}{[h(x)]}.$$

Setting $q = 1$, we obtain a formula for the number of semistandard Young tableaux:

Corollary 5.13 *We have the identity*

$$K_\lambda(n) = \prod_{x \in \lambda} \frac{n + c(x)}{h(x)}.$$

We can also use Corollary 5.2 to obtain the formula for the number of standard Young tableaux.

Corollary 5.14 (Hook length formula) *We have the identity*

$$K_\lambda = \lim_{n \to \infty} \frac{|\lambda|!}{n^{|\lambda|}} \cdot K_\lambda(n) = \frac{|\lambda|!}{\prod_{x \in \lambda} h(x)}.$$

Exercise 5.15 Compute $K_{(5,4,1,1)}(5)$ and $K_{(5,4,1,1)}$.

5.4 Problems

5.1 Show that $\begin{bmatrix} n \\ k \end{bmatrix}$ is the number of k-dimensional subspaces in an n-dimensional vector space $\mathbb{F}_q^n$ over the field of q elements.

5.2 Prove the q-analog of the Newton binomial formula:

$$(1 + tq)\left(1 + tq^2\right) \cdots (1 + tq^n) = \sum_{k=0}^{n} \begin{bmatrix} n \\ k \end{bmatrix} q^{k(k+1)/2} t^k.$$

Until now we have only considered "one-dimensional" partitions, with parts indexed by one variable. But we also can consider their two-dimensional analogs.

Definition 5.16 A collection of nonnegative integers $\lambda_{i,j}$, with $i, j \geq 1$, with the sum equal to n, is called a *plane partition* of n if it satisfies the inequalities $\lambda_{i,j} \geq \lambda_{i+1,j}$ and $\lambda_{i,j} \geq \lambda_{i,j+1}$ for each i, j.

We present "ordinary" partitions as Young diagrams. Plane partitions correspond to *three-dimensional Young diagrams*. Write the numbers $\lambda_{i,j}$ as a table and put onto the (i, j)-th box of this table a tower of $\lambda_{i,j}$ unit cubes. We get a pyramid, sometimes called a *mausoleum* made of cubes and situated "in the corner of a room" (i.e., below, behind and to the left of each cube we have a wall or another cube).

4	3	3	2	2
4	3	2	1	0
3	1	0	0	0
1	0	0	0	0

Fig. 5.4: Plane partition and the corresponding three-dimensional Young diagram

The following series of problems is devoted to the computation of the number of three-dimensional Young diagrams with bounded length, width, and height. You can find out more about three-dimensional Young diagrams and their relation to the so-called *alternating sign matrices* from the excellent (and very accessible) book [Bre99].

5.3 Show that the number of three-dimensional Young diagrams that fit into a "box" of size $a \times b \times c$ is equal to

$$\mathbb{P}(a, b, c) = \det \left(\binom{a + b + i - 1}{b + j - 1} \right)_{i,j=1}^{c}$$

$$= \det \begin{pmatrix} \binom{a+b}{b} & \binom{a+b}{b+1} & \cdots & \binom{a+b}{b+c-1} \\ \binom{a+b+1}{b} & \binom{a+b+1}{b+1} & \cdots & \binom{a+b+1}{b+c-1} \\ \vdots & \vdots & \ddots & \vdots \\ \binom{a+b+c-1}{b} & \binom{a+b+c-1}{b+1} & \cdots & \binom{a+b+c-1}{b+c-1} \end{pmatrix}.$$

5.4 Recall that the k-th falling factorial of x is defined as $x^{\underline{k}} = x(x-1) \cdots (x-k+1)$. Show that

$$\mathbb{P}(a, b, c) = \left(\prod_{j=1}^{c} \frac{(a + b + j - 1)^{\underline{b}}}{(b + j - 1)!} \right) \det \left((a + i - 1)^{\underline{j-1}} \right)_{i,j=1}^{c}.$$

5.5 Using Problem 2.1, compute the determinant $\det \left((a + i - 1)^{\underline{j-1}} \right)_{i,j=1}^{c}$ from the previous problem.

5.6 (the MacMahon formula) Prove the identities

(a)

$$\mathbb{P}(a,b,c) = \prod_{i=1}^{b}\prod_{j=1}^{c} \frac{a+i+j-1}{i+j-1};$$

(b)

$$\mathbb{P}(a,b,c) = \prod_{k=1}^{a}\prod_{i=1}^{b}\prod_{j=1}^{c} \frac{i+j+k-1}{i+j+k-2}.$$

5.7 Show that the generating function for the volumes of three-dimensional Young diagrams in the parallelepiped of size $a \times b \times c$ is equal to

$$\mathbb{P}_q(a,b,c) = \prod_{k=1}^{a}\prod_{i=1}^{b}\prod_{j=1}^{c} \frac{[i+j+k-1]}{[i+j+k-2]}.$$

5.8 Derive from this formula that the generating function for the number of arbitrary three-dimensional Young diagrams (without restrictions on their size) is equal to

$$\mathbb{P}(q) = \prod_{k=1}^{\infty} \frac{1}{\left(1-q^k\right)^k}.$$

(Compare it with Euler's product formula for partitions!)

Chapter 6
Problem Set 1

6.1 Consider the space $\Omega^{(n)} \subset \Lambda^{(n)}$ of homogeneous *supersymmetric* functions of degree n:

$$f \in \Omega^{(n)} \Leftrightarrow f(t, -t, x_3, x_4, \dots) = f(x_3, x_4, \dots).$$

Show that the functions $\{p_\lambda \mid \lambda \vdash n, \text{all } \lambda_i > 0 \text{ are odd}\}$ form a basis in $\Omega^{(n)}$.

6.2 Decompose a Young diagram λ into the union of k nested hooks with "arms" of length $a_1, \dots a_k$ and "legs" of length $b_1, \dots, b_k$. Let $s_{(a|b)} = s_{(a+1,1^b)}$ be the Schur polynomial for a hook with "arm" and "leg" of length a and b respectively. Prove the *Giambelli formula*:

$$s_\lambda = \det \left(s_{(a_i|b_j)} \right)_{i,j=1}^n .$$

Definition 6.1 Let $\lambda = (\lambda_1, \dots, \lambda_n)$ and $\mu = (\mu_1, \dots, \mu_n)$ be two partitions such that $\lambda_i \geq \mu_i$ for each i. A *skew Young diagram* λ/μ is the set of boxes obtained by removing the diagram μ from the diagram λ.

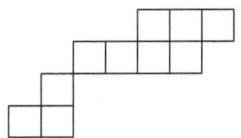

Fig. 6.1: Skew Young diagram of shape $\lambda/\mu = (7, 6, 2, 2)/(4, 2, 1, 0)$

Definition 6.2 A *skew Young tableau* is a filling of a skew diagram λ/μ with integers from 1 to n in such a way that they weakly increase along rows and strictly increase along columns. The set of all skew Young diagrams of shape λ/μ will be denoted by $\mathrm{SSYT}_{\lambda/\mu}(n)$.
The *weight* of a skew Young tableau T and the corresponding monomial $\mathbf{x}^T$ are defined similarly to the case of usual Young tableaux (Definition 3.2).

E. Smirnov, A. Tutubalina, *Symmetric Functions: A Beginner's Course*,
Moscow Lectures 10, https://doi.org/10.1007/978-3-031-50341-2_6

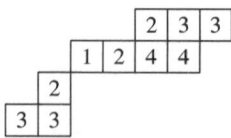

Fig. 6.2: Skew Young tableau of shape $\lambda/\mu = (7,6,2,2)/(4,2,1,0)$ and shape $(1,3,4,2)$

Definition 6.3 Define *skew Schur polynomials* as sums of monomials over all skew Young tableaux:

$$s_{\lambda/\mu}(\mathbf{x}) = \sum_{T \in \mathrm{SSYT}_{\lambda/\mu}(n)} \mathbf{x}^T.$$

6.3 Prove the Jacobi–Trudi identities for skew Schur polynomials:

(a)
$$s_{\lambda/\mu} = \det\left(h_{\lambda_i-\mu_j-i+j}\right)_{i,j=1}^{n};$$

(b)
$$s_{\lambda/\mu} = \det\left(e_{\lambda'_i-\mu'_j-i+j}\right)_{i,j=1}^{n}.$$

6.4 In this problem we define skew Schur polynomials $\tilde{s}_{\lambda/\mu}$ using the relations

$$\langle \tilde{s}_{\lambda/\mu}, s_\nu \rangle = \langle s_\lambda, s_\mu s_\nu \rangle$$

and prove that this definition is equivalent to the combinatorial one (Definition 6.3).

(a) Show that
$$\sum_\lambda \tilde{s}_{\lambda/\mu}(\mathbf{x}) s_\lambda(\mathbf{y}) = s_\mu(\mathbf{y}) \sum_\nu h_\nu(\mathbf{x}) m_\nu(\mathbf{y}).$$

(b) Show that
$$\sum_\lambda \tilde{s}_{\lambda/\mu}(\mathbf{x}) a_{\lambda+\delta}(\mathbf{y}) = \sum_{\alpha \in \mathbb{Z}_{\geq 0}^n} h_\alpha(\mathbf{x}) \sum_{w \in S_n} (-1)^w \mathbf{y}^{\alpha+w(\mu+\delta)}$$

(for any sequence $\alpha = (\alpha_1, \ldots, \alpha_n) \in \mathbb{Z}_{\geq 0}^n$, we set $h_\alpha = h_{\alpha_1} h_{\alpha_2} \ldots h_{\alpha_n}$).

(c) Obtain from this expression the Jacobi–Trudi identity for skew Schur polynomials (Problem 6.3):

$$\tilde{s}_{\lambda/\mu} = \det\left(h_{\lambda_i-\mu_j-i+j}\right)_{i,j=1}^{n}.$$

Part II
Arrays and the
Littlewood–Richardson Rule

Chapter 7
Arrays and Condensation Operations

This part of our book is devoted to a new tool for working with Young tableaux: *arrays*. They provide easy proofs of various relations for Schur polynomials. The construction of arrays was proposed by Vladimir Danilov and Gleb Koshevoy in 2005, cf. [DK05]. In our exposition we also largely follow the algebra textbook by Alexei Gorodentsev [Gor17].

7.1 Definition of Arrays

Definition 7.1 An *array* is a rectangular $m \times n$ table filled by nonnegative integers $a(i, j)$. We index the rows from bottom to top and the columns from left to right, like the points in the first quarter of the plane.

We will consider $a(i, j)$ as *quantities* of indistinguishable objects (like balls or marbles) placed into boxes of a rectangular table. Of course, a box is allowed to be empty.

Remark 7.2 Formally, arrays are integer matrices with nonnegative coefficients. But we do not want to call them matrices, for two reasons. First, matrices can be added and multiplied together, and we will never do this with arrays. On the other hand, we are going to introduce condensation operations on arrays, which do not have any meaning for matrices; they are defined by moving certain balls between neighboring boxes. The second reason is the different convention on indexing elements: matrix elements are indexed starting from the upper-left corner, with the row number first. With arrays, the convention is opposite: we put the column number first and count rows starting from below.

Counting the total number of balls in each column of an array a, we get its *column weight* $\mathrm{wt}^c(a) = \left(\mathrm{wt}^c_1(a), \ldots, \mathrm{wt}^c_m(a)\right)$; here $\mathrm{wt}^c_i(a) = \sum_{j=1}^n a(i, j)$ is the number of balls in the i-th column.

Similarly we can define the *row weight* of an array $\mathrm{wt}^r(a) = \left(\mathrm{wt}^r_1(a), \ldots, \mathrm{wt}^r_n(a)\right)$; with $\mathrm{wt}^r_j(a) = \sum_{i=1}^m a(i, j)$ being the number of balls in the j-th row.

© The Author(s), under exclusive license to Springer Nature Switzerland AG 2024
E. Smirnov, A. Tutubalina, *Symmetric Functions: A Beginner's Course*,
Moscow Lectures 10, https://doi.org/10.1007/978-3-031-50341-2_7

Arrays can be *transposed* $a \mapsto a^T$, $a^T(i,j) = a(j,i)$; another natural involutive operation is the *central symmetry* $a \mapsto a^S$, $a^S(i,j) = a(m+1-i, n+1-j)$.

Let us introduce four kinds of *condensation operations* D_j, U_j, L_i, R_i acting on arrays. Here we assume that j stands for row number, with $1 \le j \le n-1$, while i indexes columns, with $1 \le i \le m-1$. Each operation moves at most one ball to the neighboring box, taking it down, up, left, or right respectively.

7.2 Stable Matchings and Condensation Operations

Vertical operations D_j and U_j move at most one ball in the vertical direction between the rows with the numbers j and $j+1$ (we will refer to these rows as the bottom and the top one, respectively). To figure out which ball (if any) it moves, we need to establish a *stable matching* between the balls in these rows.

Definition 7.3 We are matching balls in the top row with balls in the bottom row, one by one, going from left to right along the top row. In the beginning, all balls in both rows are unmatched. Each ball in the top row is matched to the rightmost unmatched ball strictly to the left of it in the bottom row. If no such ball exists, then the ball in the top row remains unmatched. After we apply this procedure to all the balls in the top row, the remaining balls in both rows without a pair are called *free*.

Example 7.4 The matching between the rows $(1,1,2,0,3,2,1,0,2)$ (top) and $(3,2,0,0,1,4,1,0,3)$ (bottom) is shown in the figure below.

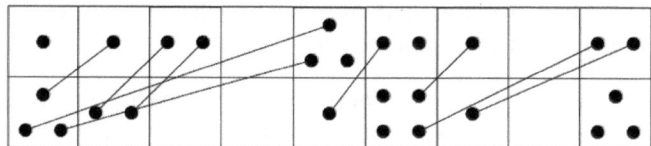

Exercise 7.5 Establish the matching between the rows

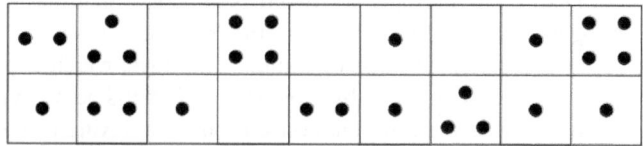

Note that every free ball in the top row is situated weakly to the left of any free ball in the bottom row (otherwise these two balls would have formed a pair).

Definition 7.6 Operation D_j brings the *rightmost free ball* in the $(j+1)$-th row to the box immediately under it, or does not do anything if all the balls in the top row are matched. Operation U_j lifts the *leftmost free ball* in the bottom row to the box immediately below it, or leaves the array unchanged if all the balls in the j-th row are matched.

An operation is said to be *inefficient* if it leaves the array unchanged, and *efficient* otherwise.

Exercise 7.7 Apply the operations D_1, D_2, U_1, U_2 to the array

0	2	1	1
3	2	1	2
1	4	0	3

Exercise 7.8 Show that the operations D_j and U_j preserve the matching between the j-th and the $(j + 1)$-th rows.

Suppose that D_j acts on an array a efficiently, bringing a ball b downwards. It becomes the leftmost free ball in the j-th row, so the operation U_j lifts it back. So we have $U_j D_j a = a$, if D_j acts effectively on a (i.e. U_j is the left inverse to D_j). Similarly we have $D_j U_j a = a$, if U_j acts efficiently on a.

We obtain a structure that resembles a group (but not quite): operations are invertible provided they are efficient. Suppose we act on a by a word $D = D_{j_1} \cdots D_{j_k}$, with every operation D_j acting efficiently (we shall say that the whole word D acts *efficiently*).

Then we can recover the initial array using the formula

$$a = U_{j_k} \cdots U_{j_2} U_{j_1} \left(D_{j_1} D_{j_2} \cdots D_{j_k} a \right).$$

Definition 7.9 *Horizontal operations* L_i and R_i are defined in a similar way: they move a ball between the i-th and $(i + 1)$-th columns of the array and turn into vertical operations if we transpose the array:

$$L_i a = \left(D_i a^T \right)^T, \quad R_i a = \left(U_i a^T \right)^T.$$

7.3 The Commutation Lemma

The following statement is essential for our study of arrays.

Lemma 7.10 *Horizontal operations L_i and R_i commute with vertical operations D_j and U_j.*

Proof Let us show that a horizontal operation L_i does not change the row matching between the j-th and the $(j + 1)$-th rows (i.e., the pairs of matched balls remain the same after applying it).

Suppose that L_i moves a ball **b** from the $(i + 1)$-th column to the i-th one. If this ball was outside the rows with the numbers j and $j + 1$, the matching will obviously remain the same. In the opposite case, consider the 2×2 square obtained by intersecting the columns i and $i + 1$ with the rows j and $j + 1$ and distinguish between two possibilities:

- The ball **b** is in the $(j + 1)$-th row.

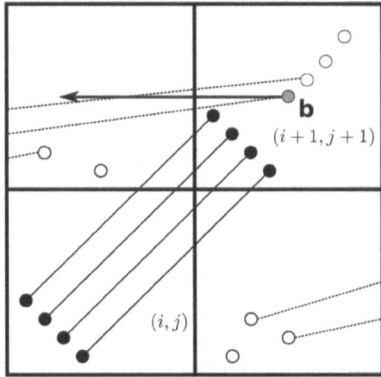

When we establish a matching between columns i and $i + 1$, the ball **b** in the box $(i + 1, j + 1)$ remains free. This means that the number of balls in the box (i, j) is strictly less than in the box $(i + 1, j + 1)$.

Now let us establish the matching between the rows. Each ball from the box (i, j) will be matched to a ball from the box $(i + 1, j + 1)$, and there will be no balls left to be matched with **b**. This means that **b** is either free (and remains free after moving it to the left) or is matched to a ball in the column with the number strictly less than i (and this matching survives when we move it to the left). The other matchings also remain unchanged.

- The ball **b** is in the j-th row.

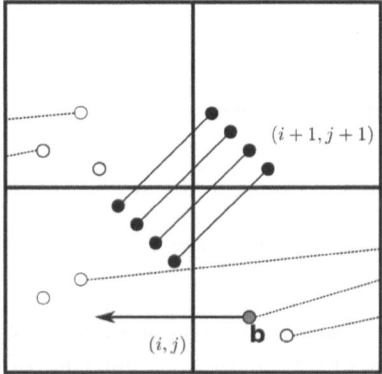

Consider the matching between columns. The ball **b** is the topmost free ball. This means that every ball in the box $(i+1, j+1)$ is matched. However, the construction of the matching suggests that we were looking (without success) for the match for the ball **b** before considering the balls in the box $(i + 1, j + 1)$. Hence the balls in this box can be matched only to the balls in the box (i, j).

Now consider the row matching. All the balls from the box $(i + 1, j + 1)$ are still matched with balls from (i, j). When we move the ball **b** to the left, it cannot become matched with a ball in $(i + 1, j + 1)$. This means that again the matching remains the same.

Summarizing, we have shown that the operations L_i (and hence R_i) preserve row matchings, and hence they preserve free balls. Similarly, the vertical operations preserve column matchings and free balls (for these matchings). This means that the horizontal and the vertical operations commute.   □

Corollary 7.11 *An operation D_j or U_j acts on an array efficiently if and only if it acts efficiently on any array obtained from it by horizontal operations.*

Proof Let a be an array, and let H be a word formed by horizontal operations. If $D_j a = a$, we have $D_j(Ha) = H(D_j a) = Ha$. On the contrary, if $D_j a \neq a$, the operation D_j moves a ball down and changes the row weight of the array. Since horizontal operations do not change the row weight, we have $Ha \neq H(D_j a) = D_j(Ha)$, and the action of D_j on Ha is efficient.   □

Obviously, the symmetric statement also holds.

7.4 Dense Arrays

Definition 7.12 An array is said to be *D-dense* (respectively U, L, R-dense) if every operation D_j (respectively U_j, L_i, R_i) acts on it identically. In other words, for each two neighboring rows $(j, j + 1)$ (respectively columns $(i, i + 1)$) all the balls in the upper row (respectively lower row, right column, left column) are matched.

Suppose we have a D-dense array a_d. The box $(1, 2)$ of this array is empty: otherwise these balls would have been free in the matching of the first two rows, and this would imply that D_1 is efficient. Similarly, the boxes $(1, 3)$ and $(2, 3)$ are also empty. Proceeding by induction, we obtain that a_d does not contain any balls above the diagonal.

Similarly, any L-dense array a_ℓ does not contain any balls below the diagonal.

Definition 7.13 An array that is D-dense and L-dense simultaneously is said to be *bidense*.

All balls in a bidense array a are situated on the diagonal. Also observe that $a(1, 1) \geq a(2, 2) \geq \cdots \geq a(k, k)$. This provides a bijection between the bidense arrays and the Young diagrams (partitions).

Applying sufficiently many efficient downward operations D_j to an array a results in a D-dense array, called the *D-condensation*, or the *downward condensation*, of a. Similarly, L-operations bring an array into an L-dense one. These condensations can be applied in several ways, and *a priori* the result can be different. But this is not the case: the result does not depend upon a choice of condensation.

Example 7.14 Consider the following two condensations of this array.

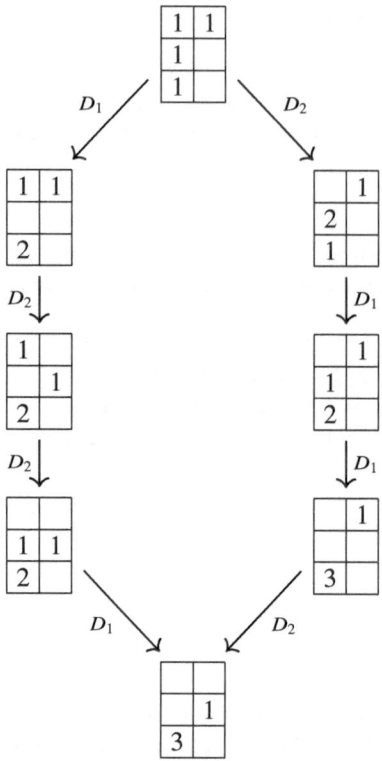

Proposition 7.15 *The condensation of an array does not depend upon a sequence of condensation operations.*

Proof First consider the case when we condensate an L-dense array a_ℓ downwards. Corollary 7.11 implies that all the intermediate arrays are L-dense, so we finally get a bidense array. The vertical operations preserve the column weight, which means that this bidense array is uniquely determined: the box (i, i) contains $\mathrm{wt}_i^c(a_\ell)$ balls.

Now let a be an arbitrary array. Let us fix a sequence of efficient operations $L = L_{i_1} \dots L_{i_k}$ condensing a to the left. Let $D = D_{j_1} \cdots D_{j_l}$ be an arbitrary sequence of downward operations providing a downward condensation of a. The array $L(Da) = D(La)$ is bidense (the equality is implied by the commutation lemma), since L preserves the D-density of Da, and D preserves the L-density of La. Moreover, L acts on Da efficiently.

As we have already seen, the array $D(La)$ does not depend upon a sequence of condensation operations D. Then $Da = L^{-1}(D(La))$ is independent of D as well.□

Definition 7.16 Let us condense an arbitrary array a downwards and to the left. The partition corresponding to the resulting bidense array is called the *shape* of a.

7.5 Problems

7.1 Compute the downward condensation of the arrays

(a)
0	1	2
1	1	4
2	1	1

(b)
1	1	1
1	1	1
1	1	1

(c)
3	3	3
2	2	2
1	1	1

(d)
1	2	3
1	2	3
1	2	3

7.2 Find the shape of an $n \times n$ array with 1 in each box.

7.3 Show that an array a is D-dense if and only if

$$a_{1,j+1} + a_{2,j+1} + \cdots + a_{i,j+1} \le a_{1,j} + a_{2,j} + \cdots + a_{i-1,j}$$

for each i, j.

7.4 Show that the central symmetry brings D_j to U_j, and vice versa (formally: $U_j(a) = \left(D_{n-j}\left(a^S\right)\right)^S$.)

7.5 Let a be an $m \times n$ array, and a^S the array obtained from it by the central symmetry.

(a) Show that the shapes of a and a^S are the same.
(b) Suppose that a is D-dense. Let $\sigma(a) = Da^S$ be the D-condensation of the array a^S. Show that the map σ is an involution of the set of D-dense arrays.

Chapter 8
Arrays and Schur Polynomials

In this chapter we discuss relations between arrays and Schur polynomials.

8.1 Dense Arrays and Young Tableaux

Let a be an array of size $m \times n$. We can transform each of its rows into a weakly increasing sequence by taking its *row scan*.

Let us read the row j from left to right. For each ball in a box (i, j), we record the letter i. We obtain a sequence

$$\underbrace{1, 1, \ldots, 1}_{a(1,j)}, \underbrace{2, 2, \ldots, 2}_{a(2,j)}, \ldots, \underbrace{m, m, \ldots, m}_{a(m,j)}$$

Write these n sequences under one another, left-adjusted, going *down*.

2	3	0	1	1			
0	4	1	0	0			
3	0	1	1	3			
1	2	0	1	0			

$\longrightarrow$

1	2	2	4				
1	1	1	3	4	5	5	5
2	2	2	2	3			
1	1	2	2	2	4	5	

The resulting table is called the *row scan* of an array. As we can see from this example, it is not necessarily a semistandard Young tableau; moreover, its shape is not necessarily a Young diagram. However, the following proposition holds true.

Proposition 8.1 *The row scan of an array a is a semistandard Young tableau if and only if a is D-dense.*

Proof By construction, numbers in the rows of the row scan weakly increase. Strict column increasing is equivalent to the following: the i-th ball in the $(j+1)$-th row of our array is located strictly to the right of the i-th ball in the j-th column. This is equivalent to the fact that each ball in the top row has a matching ball in the bottom row. $\qquad\square$

© The Author(s), under exclusive license to Springer Nature Switzerland AG 2024
E. Smirnov, A. Tutubalina, *Symmetric Functions: A Beginner's Course*,
Moscow Lectures 10, https://doi.org/10.1007/978-3-031-50341-2_8

0	0	0	0	0	1
0	0	0	0	1	0
0	0	0	2	0	2
0	1	1	0	2	0
2	0	1	3	0	1

$\longleftrightarrow$

1	1	3	4	4	4	6
2	3	5	5			
4	4					
5						
6						

Exercise 8.2 Write down the array with the following row scan:

1	1	2	2	5	6	6
2	3	4	4	6		
4	4	5				
5	6					

Show that it is indeed D-dense.

Note that the column weight of a D-dense array coincides with the weight of its row scan Young tableau.

8.2 Dense Arrays and Yamanouchi Texts

Let a be an L-dense array of size $m \times n$. We can consider its *column scan* (i.e., the row scan of the transposed array a^T); it is a Young tableau with at most m rows, filled with numbers not exceeding n.

However, it is also instructive to state the L-density condition in terms of row scans.

The L-density condition means that each ball **b** in the $(i + 1)$-th column has the matching ball **b′** in the i-th column (by definition, **b′** is situated below **b**). This means that for each occurrence of $i + 1$ in the row reading we have an occurrence of i in the upper row. This condition can be restated as follows.

Definition 8.3 Read the row scan of an array row by row, top to bottom, reading each row *from right to left*. If in every initial subword of this sequence the number of 1s is greater than or equal to the number of 2s, the number of 2s is greater than or equal to the number of 3s, and so on, such a row scan is called *a Yamanouchi text*.

For example, the left diagram is a Yamanouchi text, while the right one is not.

1	1		
1	2	2	
1	2	3	3
3			

1	1	1	1
2	2	2	
2	3	3	
1	2		

Hence, an array is L-dense if and only if its row scan is Yamanouchi.

Exercise 8.4 Transpose the array that you have obtained in Exercise 8.2 and write down its row scan. Show that it is indeed Yamanouchi.

8.3 The Fiber Product Theorem

Now we will need one general set-theoretic definition.

Definition 8.5 Consider three sets X, Y, and Z and two maps $\varphi: X \to Z$ and $\psi: Y \to Z$. Then the set

$$X \underset{Z}{\times} Y := \{(x, y) \mid \varphi(x) = \psi(y)\} = \bigsqcup_{z \in Z} \varphi^{-1}(z) \times \psi^{-1}(z)$$

is called the *fiber product* of X and Y over Z.

For example, if $Z = \{z\}$ is a singleton, we have $\varphi^{-1}(z) = X$ and $\psi^{-1}(z) = Y$, so $X \underset{Z}{\times} Y$ is just the Cartesian product of the sets $X \times Y$.

This can also be viewed in the following way. Let $\pi_X: (x, y) \mapsto x$ and $\pi_Y: (x, y) \mapsto y$ be the first and the second projections. Then the following diagram (called a *Cartesian square*) is commutative:

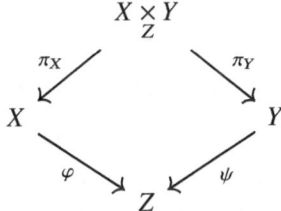

The fiber product is a universal object in the following sense. Consider an arbitrary commutative square: $\xi: M \to X$, $\zeta: M \to Y$ and $\varphi \circ \xi = \phi \circ \zeta$. Then these maps factor through $X \underset{Z}{\times} Y$: there is a unique morphism $\alpha: M \to X \underset{Z}{\times} Y$ such that $\pi_X \circ \alpha = \xi$ and $\pi_y \circ \alpha = \zeta$.

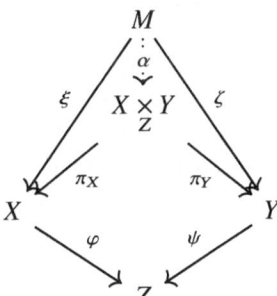

Exercise 8.6 Check this.

We can apply this definition to arrays. Denote the set of all arrays of size $m \times n$ by $\mathcal{M}$, and denote the sets of L-dense and D-dense arrays by $\mathcal{L}$ and $\mathcal{D}$ respectively. Let $\mathcal{B}$ be the set of bidense arrays, and let D and L stand for the downward and leftward condensation. Consider the following square; according to Lemma 7.10, it is commutative.

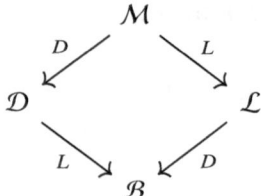

It factors through the fiber product $\mathcal{D} \underset{\mathcal{B}}{\times} \mathcal{L}$:

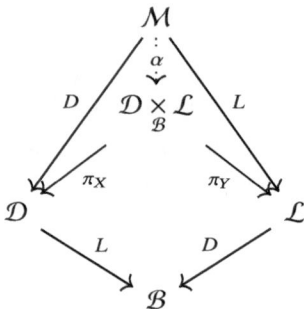

It turns out that the map $\alpha: a \mapsto (Da, La)$ is a bijection.

Theorem 8.7 *The set* $\mathcal{M}$ *of all arrays is the fiber product* $\mathcal{D} \underset{\mathcal{B}}{\times} \mathcal{L}$ *of the sets of D-dense and L-dense arrays over the set of bidense arrays.*

In other words, for any L-dense array a_ℓ *and D-dense array* a_d *such that* $Da_\ell = La_d$ *there exists a unique array* a *satisfying* $a_\ell = La$ *and* $a_d = Da$.

Proof First let us prove the injectivity of α. Let a and a' be two arrays satisfying $La = La'$ and $Da = Da'$. Consider a word Λ that condensates $Da = Da'$ to the left efficiently. Then Lemma 7.10 implies that Λ provides an efficient L-condensation of both a and a', while $\Lambda a = La = La' = \Lambda a'$, which means that $a = \Lambda^{-1}La = \Lambda^{-1}La' = a'$.

Now let us show that α is surjective. Let a_ℓ and a_d be L-dense and D-dense arrays respectively that satisfy $Da_\ell = La_d$. Consider a word Λ providing an efficient L-condensation of a_d; let the resulting array be denoted by La_d. The inverse word Λ^{-1} acts efficiently on $La_d = Da_\ell$, and hence on a_ℓ. Then $a = \Lambda^{-1}a_\ell$ is the desired preimage of the pair (a_d, a_ℓ). Indeed, we have $La = a_\ell$ and $Da = D\Lambda^{-1}a_\ell = \Lambda^{-1}Da_\ell = \Lambda^{-1}La_d = a_d$. $\qquad\square$

8.4 The Robinson–Schensted–Knuth Correspondence

Theorem 8.7 implies several important statements about Young tableaux and Schur polynomials.

Consider a square array a of size $n \times n$ with exactly one ball in each column and each row. It can be viewed as the *graph* of a certain *permutation* $w \in S_n$, with the balls situated in boxes $(i, w(i))$.

Take the D-condensation of a. The resulting array Da still has exactly one ball in each column, so its row reading is a standard Young tableau T_1. Similarly we can take the L-condensation of a, and the column reading of the resulting array La is a standard tableau T_2 of the same shape as T_1.

This means that each permutation corresponds to a pair of standard Young tableaux of the same shape. Theorem 8.7 states that this correspondence is a bijection.

Definition 8.8 The bijection between permutations $w \in S_n$ and pairs of standard tableaux of the same shape (T_1, T_2) is called the *Robinson–Schensted–Knuth correspondence* (or just RSK-correspondence).[1]

Exercise 8.9 (a) For each permutation $w \in S_3$, find the corresponding pair of Young tableaux.

(b) Let $w \in S_n$ correspond to the pair of tableaux (T_1, T_2). What is the pair of tableaux corresponding to w^{-1}?

Corollary 8.10 *The Kostka numbers satisfy the equality*

$$\sum_{\lambda \vdash n} K_\lambda^2 = n!$$

Remark 8.11 This formula has a nice representation-theoretic interpretation. Indeed, irreducible representations of permutation group S_n are indexed by partitions $\lambda \vdash n$, and the dimensions of these representations are equal to K_λ. So we recover Burnside's formula stating that for a finite group, the sum of squares of dimensions of all irreducible representations is equal to its order.

8.5 One More Proof of the Cauchy Formula

Consider an arbitrary array a of size $m \times n$. Let $\mathbf{x} = (x_1, \ldots, x_m)$, $\mathbf{y} = (y_1, \ldots, y_n)$. Let us assign the weight $x_i y_j$ to each ball in the box (i, j) and take the product of these weights over all balls in a; this monomial will be called the *weight* of a. Now take the sum of these weights for all possible arrays. Each box (i, j) can contain an arbitrary number of balls, so this sum is equal to the Cauchy product:

$$\prod_{i=1}^{m} \prod_{j=1}^{n} \left(1 + x_i y_j + x_i^2 y_j^2 + \cdots \right) = \prod_{i,j} \frac{1}{1 - x_i y_j}.$$

This expression can also be computed in a different way. Consider all D-dense arrays of shape λ; the sum of their $\mathbf{x}$-weights equals $s_\lambda(\mathbf{x})$. Similarly the sum of $\mathbf{y}$-weights of all possible L-dense arrays of shape λ is equal to $s_\lambda(\mathbf{y})$.

---

[1] The algorithm transforming a permutation into a pair of Young tableaux was discovered independently by Robinson [Rob38] and Schensted [Sch61]. Instead of arrays they used the so-called *row insertion* on Young tableaux. Donald Knuth [Knu70] extended this algorithm for *generalized permutations* and used it to prove the Cauchy formula.

Let us now use Theorem 8.7 and the following fact: the **x**-weight remains the same under the action of D-condensation, while L-condensation does not change the **y**-weight. This means that the sum of weights over all arrays of shape λ equals $s_\lambda(\mathbf{x})s_\lambda(\mathbf{y})$. Taking the sum over λ, we obtain the Cauchy formula:

$$\sum_\lambda s_\lambda(\mathbf{x})s_\lambda(\mathbf{y}) = \prod_{i,j} \frac{1}{1 - x_i y_j}.$$

8.6 The Schur Formula

Now consider all possible *symmetric* arrays. To each ball in box (i, j) we assign the weight $\sqrt{z_i z_j} = x_i y_j|_{\mathbf{x}=\mathbf{y}=\sqrt{\mathbf{z}}}$. Let a be a symmetric array of shape λ. As we know, D-condensation preserves the **x**-weight, so the **x**-weight of a is a monomial occurring in $s_\lambda(\mathbf{x})$. Similarly, the **y**-weight of a is *the same* monomial in $s_\lambda(\mathbf{y})$. Substituting $\mathbf{x} = \mathbf{y} = \sqrt{\mathbf{z}}$, we see that the weight of a is a monomial in $s_\lambda(\mathbf{z})$, and all monomials occur in such a way.

Taking the sum of arrays' weights first over all symmetric arrays of a given shape λ, and then over all λ, we obtain $\sum_\lambda s_\lambda(\mathbf{z})$. On the other hand, this sum is equal to

$$\prod_i \left(1 + z_i + z_i^2 + \cdots\right) \cdot \prod_{i<j} \left(1 + z_i z_j + z_i^2 z_j^2 + \dots\right) = \prod_i \frac{1}{1 - z_i} \cdot \prod_{i<j} \frac{1}{1 - z_i z_j}.$$

We get *the Schur formula*:

$$\sum_\lambda s_\lambda(\mathbf{z}) = \prod_i \frac{1}{1 - z_i} \cdot \prod_{i<j} \frac{1}{1 - z_i z_j}.$$

8.7 Problems

8.1 Apply the RSK-correspondence to a given permutation $w \in S_n$ and find the pair of standard Young tableaux:

(a) $w = \overline{n, n - 1, \ldots, 2, 1} \in S_n$;
(b) $w = \overline{(i - 1 \leftrightarrow i)} \in S_n$;
(c) $w = \overline{471932586} \in S_9$.

8.2 Apply the RSK-correspondence to the following pairs of standard Young tableaux and recover the corresponding permutation:

(a)

1	3	4	5	6	7	9
2
8

and

1	4	5	6	7	8	9
2
3

(b)

1	3	5	8
2	4	7	
6	9		

and

1	2	4	7
3	6	9	
5	8		

8.3 Prove the following identities on Kostka numbers:

(a)
$$\sum_{\lambda \vdash n} K_\lambda = \left|\{\sigma \in S_n \mid \sigma^2 = id\}\right|;$$

(b)
$$\sum_{\lambda \vdash n} K_\lambda \cdot K_\lambda(m) = m^n;$$

(c)
$$\sum_{\lambda \vdash n} K_\lambda(m) \cdot K_\lambda(k) = \binom{mk + n - 1}{n}.$$

8.4 (a) Let a be a symmetric array of size $n \times n$ with the numbers $a(1, 1)$, $a(2, 2)$, $\ldots, a(n, n)$ on the diagonal. Consider the LD-bicondensation of a, denoted by LDa; let $\lambda_1, \ldots, \lambda_n$ be the numbers on its diagonal. Show that the sequences λ_i and $a(i, i)$ have the same number of odd terms.

(b) A partition λ is said to be *even* if all its parts λ_i are even. Using arrays, prove the identity
$$\sum_{\lambda \text{ even}} s_\lambda(\mathbf{x}) = \prod_i \frac{1}{1 - x_i^2} \cdot \prod_{i<j} \frac{1}{1 - x_i x_j}.$$

(c) Show that
$$\left(\sum_{\lambda \text{ even}} s_\lambda(\mathbf{x})\right)\left(\sum_{k \geq 0} e_k(\mathbf{x})\right) = \left(\sum_\lambda s_\lambda(\mathbf{x})\right).$$

As a consequence, obtain the following identity:
$$\sum_{\lambda' \text{ even}} s_\lambda(\mathbf{x}) = \prod_{i<j} \frac{1}{1 - x_i x_j}.$$

Chapter 9
The Littlewood–Richardson Rule

In this chapter we will obtain the rule for expanding products of Schur polynomials as linear combinations of other Schur polynomials. In other words, we will compute the *structure constants* $c^\nu_{\lambda\mu}$, given by the equality

$$s_\lambda s_\mu = \sum_\nu c^\nu_{\lambda\mu} s_\nu.$$

9.1 *DU*-Sets

Start with an arbitrary set of arrays and apply to them all possible sequences of vertical operations D_j and U_j. The resulting collection of arrays is called a *DU-set*.

Definition 9.1 A set of arrays that is stable under all vertical operations is called a *DU-set*. A *DU*-set is called a *DU-orbit* if the action of operations D_j and U_j is transitive.

Every *DU*-orbit has a unique *D*-dense array a_d; we shall call it the *bottom array* of this orbit. All other arrays in the *DU*-orbit are obtained from the bottom array by applying efficient operations U_j.

Exercise 9.2 Show that:

(a) the union, the intersection and the difference of *DU*-sets are again *DU*-sets;
(b) every *DU*-set is a disjoint union of *DU*-orbits;
(c) horizontal operations L_i and R_i take *DU*-orbits into *DU*-orbits.

Definition 9.3 If the bottom array of a *DU*-orbit O_λ is a bidense array of shape λ, then such an orbit is said to be *standard*.

All arrays in a standard orbit O_λ are *L*-dense.
The left condensation operation L defines a bijection of any *DU*-orbit O with a standard orbit O_λ. Here λ is the shape of the bottom array in O (we shall say that O has *type* λ).

E. Smirnov, A. Tutubalina, *Symmetric Functions: A Beginner's Course*,
Moscow Lectures 10, https://doi.org/10.1007/978-3-031-50341-2_9

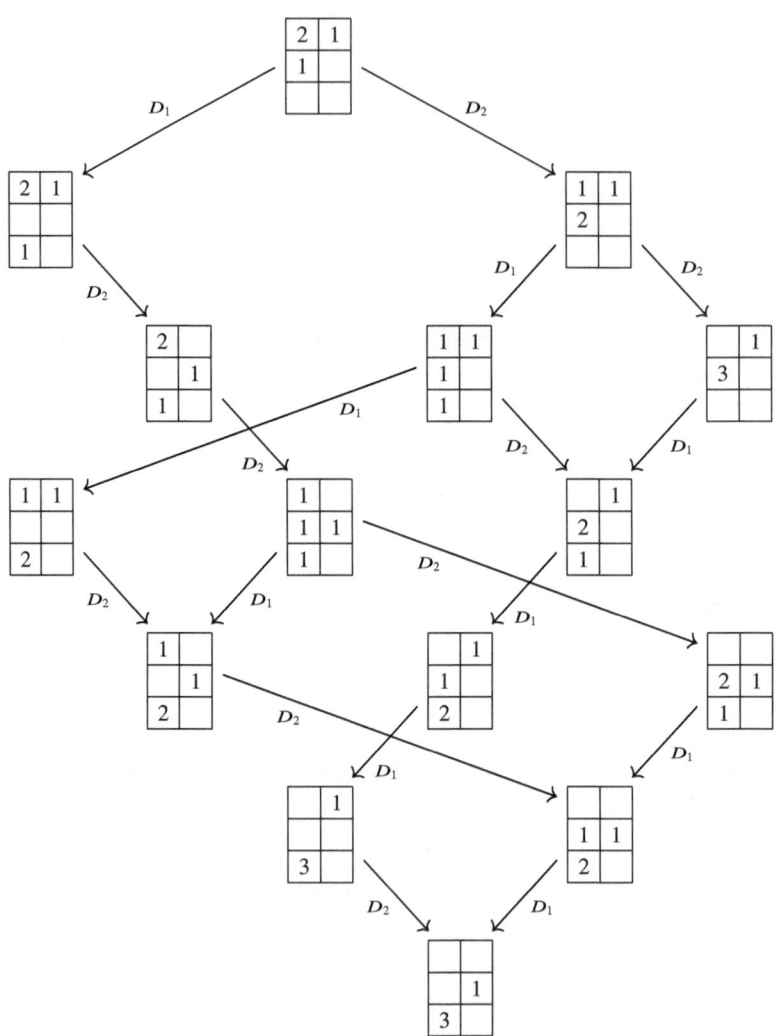

Fig. 9.1: Standard orbit $O_{(3,1)}$

$$s_{(3,1)}(x_1, x_2, x_3) =$$
$$x_1^3 x_2 + x_1^3 x_3 + x_1^2 x_2^2 + 2x_1^2 x_2 x_3 + x_1 x_2^3 + x_1^2 x_3^2 + 2x_1 x_2^2 x_3 + 2x_1 x_2 x_3^2 + x_2^3 x_3 + x_1 x_3^3 + x_2^2 x_3^2 + x_2 x_3^3$$

9.2 *DU*-Orbits and Schur Polynomials

To each ball in the j-th row of an array a we can assign the weight x_j. The product $\mathbf{x}^a$ of these weights over all balls in a will be called the *weight* of a.

The standard orbit O_λ consists of all L-dense arrays of shape λ. The column reading of each such array is a semistandard Young tableau. Hence the sum of weights over all arrays from O_λ equals the Schur polynomial:

$$\sum_{a \in O_\lambda} \mathbf{x}^a = s_\lambda(\mathbf{x}).$$

Moreover, horizontal operations do not change the row weight of an array, so

$$\sum_{a \in O} \mathbf{x}^a = s_\lambda(\mathbf{x}),$$

where O is an arbitrary DU-orbit of type λ.

Consider an arbitrary DU-set M. It is a disjoint union of certain DU-orbits; this fact allows us to expand the sum of weights of arrays from M in the basis of Schur polynomials:

$$\sum_{a \in M} \mathbf{x}^a = \sum_{\lambda} c_M^\lambda s_\lambda(\mathbf{x}),$$

where c_M^λ is the number of DU-orbits of type λ in this union. The number c_M^λ can be computed as the number of D-dense arrays of shape λ in the set M.

9.3 Products of Orbits

Let us use this idea to expand the product $s_\lambda s_\mu$ in the basis s_ν. We need a DU-set M such that

$$\sum_{a \in M} \mathbf{x}^a = s_\lambda s_\mu.$$

Let us take the "product" of DU-orbits O_λ and O_μ as follows. We take all arbitrary arrays $b \in O_\mu$ (of size $m \times n$) and write them to the right of all possible arrays $a \in O_\lambda$ (also of size $m \times n$). By definition, all such "thick" orbits ab (of size $2m \times n$) form the product of orbits $O_\lambda \otimes O_\mu$. It is clear that

$$|O_\lambda| \cdot |O_\mu| = |O_\lambda \otimes O_\mu|$$

and

$$s_\lambda(\mathbf{x}) \cdot s_\mu(\mathbf{x}) = \left(\sum_{a \in O_\lambda} \mathbf{x}^a \right) \left(\sum_{b \in O_\mu} \mathbf{x}^b \right) = \sum_{\substack{a \in O_\lambda \\ b \in O_\mu}} \mathbf{x}^a \cdot \mathbf{x}^b = \sum_{ab \in O_\lambda \otimes O_\mu} \mathbf{x}^{ab}.$$

One thing that remains unclear is that the set $O_\lambda \otimes O_\mu$ is a DU-set. Let us prove this.

Lemma 9.4 *Let D_j act efficiently on $ab \in O_\lambda \otimes O_\mu$. Then $D_j(ab)$ also belongs to $O_\lambda \otimes O_\mu$.*

Proof The operation D_j brings down the rightmost free ball in the $(j + 1)$-th row of the array ab. If this ball belongs to the array a, then after removing the array b it still remains the rightmost free ball. In this case we have $D_j a = a' \in O_\lambda$, and $D_j(ab) = a'b$ is indeed in $O_\lambda \otimes O_\mu$.

Now suppose that the ball **b** moved by the operation D_j belongs to the array b. All the balls to the right of **b** are matched, and the balls they are matched with also belong to b (we exhausted all possible pairs in a before reaching **b**). If we remove the array a, all these pairs will remain. This means that **b** is again the rightmost free ball, and the action of D_j on b moves it down.

So in this case we have $D_j b = b' \in O_\mu$ and $D_j(ab) = ab' \in O_\lambda \otimes O_\mu$.   □

The same is true of the operations U_j. This means that $O_\lambda \otimes O_\mu$ is indeed a DU-set.

9.4 The Littlewood–Richardson Rule

It remains to find the D-dense arrays in the product of two standard orbits. Let $ab \in O_\lambda \otimes O_\mu$ be a D-dense array of shape ν. The array $a \in O_\lambda$ itself is D-dense; moreover, it is L-dense. This means that a is a bidense array of shape λ.

The row scan of ab is a Young tableau of shape ν. Inside ν we have a diagram of shape λ with a "canonical" filling: it has 1s in the first row, 2s in the second, etc., while the skew diagram ν/λ is somehow filled by the numbers in the range from $m + 1$ to $2m$. Let us remove the diagram λ and decrease every number in ν/λ by m.

We get a "skew row scan" of array b. It is a *skew Young tableau*: its entries strictly increase along the columns and weakly increase along the rows.

Now recall that the array $b \in O_\mu$ is L-dense, with shape μ. This means that its row scan is a Yamanouchi text with μ_1 1s, μ_2 2s, etc.

Summarizing, we obtain the following theorem:

Theorem 9.5 (The Littlewood–Richardson rule) [1] *In the decomposition*

$$s_\lambda s_\mu = \sum_\nu c_{\lambda\mu}^\nu s_\nu$$

the sum is taken over all Young diagrams ν obtained from λ by adding $|\mu|$ boxes. The coefficient $c_{\lambda\mu}^\nu$ is equal to the number of fillings of these boxes by μ_1 1s, μ_2 2's etc, in such a way that the result is a skew Young tableau of shape ν/λ that is simultaneously a Yamanouchi text.

In other words, the numbers in the skew diagram ν/λ must weakly increase along rows and strictly increase along rows. For the reading word of this diagram from right to left, from top to bottom, in every initial subword of this sequence the number of 2s does not exceed the number of 1s, the number of 3s does not exceed the number of 2s, and so on.

Remark 9.6 Note that the Littlewood–Richardson rule does *not* imply the equality $c_{\lambda\mu}^\nu = c_{\mu\lambda}^\nu$, which is obvious from the definition of the Littlewood–Richardson coefficients (the product of polynomials is commutative!). There exist other combinatorial interpretations of the coefficients $c_{\lambda\mu}^\nu$, and some of them imply this symmetry as well as other symmetries. One remarkable example of a "more symmetric" description of the Littlewood–Richardson coefficients is given by the *Knutson–Tao puzzle rule* (cf., for example, [Egg19]).

9.5 Problems

9.1 Using the Littlewood–Richardson rule, compute the product of Schur polynomials:

    (a) $s_{(2,1)} \cdot s_{(1,1)}$;
    (b) $s_{(2,1)} \cdot s_{(2,1)}$;
    (c) $s_{(2,2)} \cdot s_{(2,1)}$;
    (d) $s_{(3,1)} \cdot s_{(2,1)}$.

9.2 Derive the Pieri formulas from the Littlewood–Richardson rule.

9.3 Let λ be decomposed as a union of hooks $\gamma_1, \gamma_2, \ldots, \gamma_m$, as shown in the figure below.

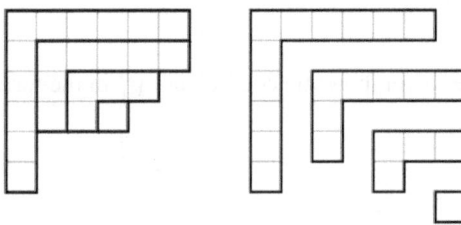

---

[1] This rule was first stated (without proof) by Littlewood and Richardson in [LR34]. The first complete proofs are due to Thomas [Tho74] and Schützenberger [Sch77].

Find the coefficient in front of s_λ in the expansion of $s_{\gamma_1} s_{\gamma_2} \ldots s_{\gamma_m}$ in the basis of Schur polynomials.

9.4 (a) Prove that the power sums p_k can be computed as alternating sums of Schur polynomials for hook-shaped Young diagrams:

$$p_k = s_{(k)} - s_{(k-1,1)} + s_{(k-2,1^2)} - s_{(k-3,1^3)} + \cdots + (-1)^k s_{(1^k)}$$

(b) (the Murnaghan–Nakayama rule) Prove the formula for the product of a Schur polynomial and a power sum:

$$p_k s_\lambda = \sum_\mu \pm s_\mu,$$

where the sum is taken over the diagrams μ obtained from λ by adding a "ribbon" of k boxes along the edge (a ribbon should be edge-connected and should not contain 2×2 squares). The corresponding Schur polynomial is taken with the plus sign if the height of the ribbon added is odd, and with the minus sign if it is even.

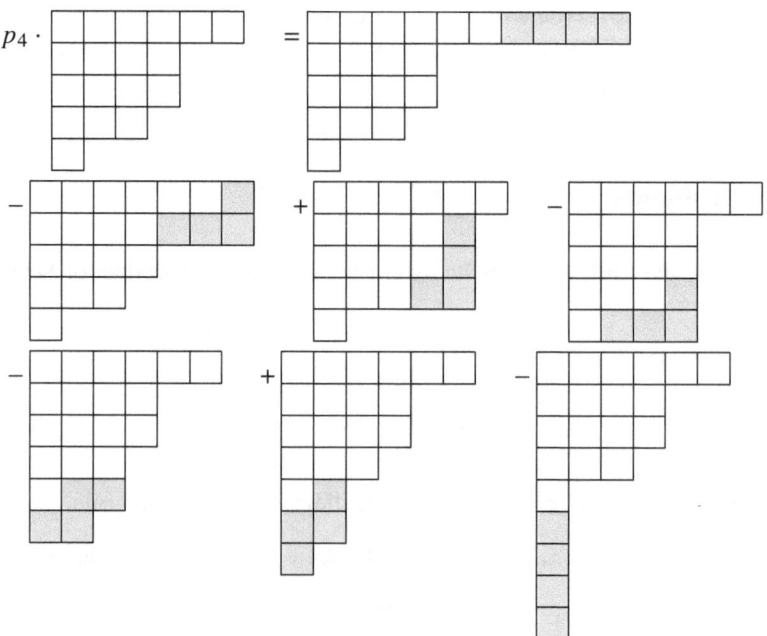

(c) Find a rule to compute the coefficients χ_λ^μ in the formula

$$p_\lambda = \sum_\mu \chi_\lambda^\mu s_\mu.$$

(d) Compute $\chi_{(5,4,4,2,1)}^{(5,5,4,2)}$.

Chapter 10
Problem Set 2

Definition 10.1 The *Schensted insertion* is an operation producing from a semistandard Young tableau T filled with numbers t_{ij} and a positive integer x a new tableau $T' = (T \leftarrow x)$ consisting of $|T| + 1$ boxes, according to the following rules:

- if x is greater than or equal to all numbers in the first row of T, then T' is obtained from T by adding an x-labeled box into the first row;
- if x is less than the maximal element in the first row: let y be the leftmost element of the first row such that $x < y$. Replace y by x in the first row (we shall say that x "bumps out" y from the row);
- insert y into the second row according to the same rule: if all the numbers in the second row do not exceed y, we add y to the end of the row; otherwise replace by y the leftmost number that is greater than y, bumping it out;
- insert the obtained number into the third row, and so on.

Example 10.2 Consider the following example of insertion of a number into a Young tableau. The boxes where the value is changed (as a result of adding or bumping out) are shaded.

$$
\begin{array}{|c|c|c|c|}
\hline
1 & 1 & 3 & 5 \\
\hline
2 & 3 & 6 \\
\cline{1-3}
3 & 5 \\
\cline{1-2}
\end{array}
\leftarrow 2
\quad
\begin{array}{|c|c|c|c|}
\hline
1 & 1 & 2 & 5 \\
\hline
2 & 3 & 6 \\
\cline{1-3}
3 & 5 \\
\cline{1-2}
\end{array}
\leftarrow 3
$$

$$
\begin{array}{|c|c|c|c|}
\hline
1 & 1 & 2 & 5 \\
\hline
2 & 3 & 3 \\
\cline{1-3}
3 & 5 \\
\cline{1-2}
\end{array}
\leftarrow 6
\quad
\begin{array}{|c|c|c|c|}
\hline
1 & 1 & 2 & 5 \\
\hline
2 & 3 & 3 \\
\cline{1-3}
3 & 5 & 6 \\
\cline{1-3}
\end{array}
$$

10.1 (a) Show that the result of insertion is again a semistandard Young tableau.
 (b) Show that the Schensted insertion is invertible: given a tableau T' and its corner box, we can uniquely recover the tableau T and the number x, such that $T' = (T \leftarrow x)$.

© The Author(s), under exclusive license to Springer Nature Switzerland AG 2024
E. Smirnov, A. Tutubalina, *Symmetric Functions: A Beginner's Course*,
Moscow Lectures 10, https://doi.org/10.1007/978-3-031-50341-2_10

10.2 (Robinson–Schensted correspondence) Let $(x_i) = (x_1, \ldots, x_m)$ be an arbitrary sequence of positive integers not exceeding n, and let

$$P = (\ldots ((\varnothing \leftarrow x_1) \leftarrow x_2) \leftarrow \ldots) \leftarrow x_m$$

be the Young tableau obtained by insertion of these numbers one by one. Consider a *standard* tableau Q of the same shape, with i situated in the box added to P on the i-th step (it is called the *counting tableau*). We put into correspondence to (x_i) the pair of tableaux (P, Q).

(a) Show that this map is a bijection between the set of maps from $[1, m]$ to $[1, n]$ and the set of pairs of tableaux (P, Q) of the same shape filled by integers not exceeding n, where P is semistandard, and Q is standard.

(b) Construct a bijection between the set of permutations S_n and the set of pairs of *standard* tableaux of the same shape consisting of n boxes.

Definition 10.3 A *quasipermutation* is a two-row matrix with positive integer entries

$$\begin{pmatrix} i_1 & i_2 & \ldots & i_m \\ j_1 & j_2 & \ldots & j_m \end{pmatrix},$$

such that $i_1 \leq i_2 \leq \cdots \leq i_m$, and $i_r = i_{r+1}$ implies $j_r \leq j_{r+1}$. Ordinary permutations and maps from $[1, m]$ to $[1, n]$ are particular examples of quasipermutations; the latter maps correspond to the case $(i_1, \ldots, i_m) = (1, \ldots, m)$, $j_r \leq n$.

10.3 (Knuth correspondence) Describe a bijection between the set of quasipermutations and pairs of semistandard tableaux of the same shape, such that a quasipermutation is mapped to a pair of tableaux filled by $(j_1, \ldots, j_m)$ and $(i_1, \ldots, i_m)$ respectively, in a way generalizing the correspondences from the previous exercise.

10.4 Derive the identities stated in Exercise 8.3 from the previous problem.

10.5 For a given quasipermutation, consider a monomial in two sets of variables $\mathbf{x}$, $\mathbf{y}$, equal to $(x_{i_1} y_{j_1}) \ldots (x_{i_m} y_{j_m})$. Write the generating function for quasipermutations in two different ways and obtain the Cauchy identity:

$$\prod_{i,j} (1 - x_i y_j)^{-1} = \sum_{\lambda} s_\lambda(\mathbf{x}) s_\lambda(\mathbf{y}).$$

Definition 10.4 Define the *plactic* [1] *monoid* as the set of words in an n-letter alphabet $\{1, \ldots, n\}$, modulo the *Knuth equivalence relations*:

$$xzy \sim zxy \text{ for } x \leq y < z; \qquad yzx \sim yxz \text{ for } x < y \leq z. \qquad (10.1)$$

The product in this monoid is given by concatenation, i.e. by writing the factors consecutively.

---

[1] From the Greek $\pi \lambda \alpha \xi \iota \varsigma$ — "plate, flat stone".

Next, assign to a semistandard Young tableau T the word $m(T)$ obtained by reading its rows *from left to right, from bottom to top*. Such a word is called a *tableau word*. It is clear that the initial Young tableau is uniquely recovered from the tableau word.

10.6 (a) Let $T' = T \leftarrow x$. Show that $m(T') \sim m(T)x$.
       (b) Show that each Knuth equivalence class contains at least one tableau word.

10.7 Let w be an arbitrary word, i.e. a sequence of letters $1, \dots, n$. Denote by $\ell_i(w)$ the maximal total length of i nonintersecting nondecreasing subsequences in w.

       (a) Let $w \sim u$. Show that $\ell_i(w) = \ell_i(u)$ for each i.
       (b) Let $w = m(T)$, where T is a semistandard Young tableau of shape λ. Show that $\ell_i = \lambda_1 + \cdots + \lambda_i$.
       (c) Show that each Knuth equivalence class contains exactly one tableau word. This establishes a bijection between semistandard Young tableaux and Knuth equivalence classes.
       (d) Prove the *Erdős–Szekeres theorem*: every number sequence of length $pq+1$ has either a nondecreasing subsequence of length $p + 1$ or a decreasing subsequence of length $q + 1$.

Schubert Polynomials and Pipe Dreams

Chapter 11
The Symmetric Group

This part will be devoted to the study of *partially symmetric functions*: functions that are symmetric in some variables, while not necessarily being symmetric in the others. More precisely, we will be working with *Schubert polynomials*; as we will see, they generalize the notion of Schur polynomials.

Schubert polynomials are closely related with the symmetric group S_n. In this preliminary chapter we consider properties of this group.

11.1 Permutations, Systems of Generators, Parity

Let us recall some standard facts about permutations.

A bijective map $w: \{1, 2, \ldots, n\} \rightarrow \{1, 2, \ldots, n\}$ is called a *permutation*. The set of permutations of n letters forms the *symmetric group* S_n, with the group operation given by the composition of permutations. We will usually write permutations using the *one-line notation*: $w = \overline{w(1) \ldots w(n)}$.

Here are some special permutations:

- *The longest permutation* $w_{0,n} = \overline{n, n-1, \ldots, 2, 1}$ reverses the order of numbers $1, \ldots, n$.
- *Transpositions* $(i \leftrightarrow j)$ interchange two entries i and j and map all the remaining entries to themselves.
- *Cycles* $(i_1, i_2, \ldots, i_k)$ make a cyclic shift of some given subset of entries: $i_j \mapsto i_{j+1}, i_k \mapsto i_1$.

It is well known that every permutation can be decomposed into a product of transpositions. It also admits a presentation as a product of disjoint cycles.

Exercise 11.1 Show that the group S_n is generated by:

(a) transpositions $(1 \leftrightarrow i)$, where $2 \leq i \leq n$;
(b) simple transpositions $(i \leftrightarrow i + 1)$, where $1 \leq i \leq n - 1$;
(c) a transposition $(1 \leftrightarrow 2)$ and a long cycle $(1, 2, \ldots, n)$.

© The Author(s), under exclusive license to Springer Nature Switzerland AG 2024
E. Smirnov, A. Tutubalina, *Symmetric Functions: A Beginner's Course*,
Moscow Lectures 10, https://doi.org/10.1007/978-3-031-50341-2_11

Decompose a permutation w into a product of transpositions $(i_1 \leftrightarrow j_1) \circ \cdots \circ (i_k \leftrightarrow j_k)$. For each such product, the parity of the number of transpositions is the same. It will be called the *parity*, or the *sign* of permutation w; notation: $\mathrm{sgn}(w)$ or $(-1)^w$.

11.2 Relations in the Symmetric Group

In the previous section we have seen several systems of generators for the symmetric group. We will mostly use one of them, formed by simple transpositions.

Definition 11.2 Permutations $s_i = (i \leftrightarrow i + 1)$ are called *simple transpositions*.

In Exercise 11.1 you have shown that the simple transpositions $s_1, \ldots, s_{n-1}$ generate the whole group $\mathcal{S}_n$. They satisfy the following relations:

- $s_i^2 = e$;
- $s_i s_j = s_j s_i$ for $|i - j| \geq 2$ (*far commutativity*);
- $s_i s_{i+1} s_i = s_{i+1} s_i s_{i+1}$ (*braid relation*)

Definition 11.3 A sequence of simple transpositions $Q = \left(s_{i_1}, s_{i_2}, \ldots, s_{i_\ell}\right)$ is called a *word* of permutation $w \in \mathcal{S}_n$ if $w = s_{i_1} s_{i_2} \ldots s_{i_\ell}$.

Recall that simple transpositions (as well as any other permutations and, even more generally, as functions) are applied from right to left.

A decomposition of a permutation into a product of simple transpositions can be presented by the following diagram. Let $w = s_{i_1} s_{i_2} \ldots s_{i_\ell}$. We will start with n parallel strands and intertwine them, interchanging the i-th and $(i + 1)$-th strands each time we see a letter s_i. The letters in a word are read from left to right, and the strands also go from left to right. Let us index their left and right ends by the numbers $1, \ldots, n$. Then every strand joins i from the right with $w(i)$ from the left. This object is called a *wiring diagram*.

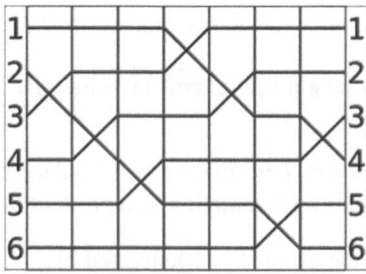

Fig. 11.1: The wiring diagram for $s_2 s_3 s_4 s_1 s_2 s_5 s_3 = \overline{345162}$.

For each permutation we can write arbitrarily many decompositions, just by inserting pairs of consecutive identical letters $s_i s_i$ into any word. This does not add much, so out of all possible decompositions we will be mostly interested in the shortest ones.

Definition 11.4 The *length* $\ell(w)$ of a permutation $w \in S_n$ is the minimal length of a word representing w. Words of length $\ell(w)$ are said to be *reduced*.

There is a different definition: the length of a permutation w is defined as the number of *inversions* $\text{inv}(w)$. An inversion is a pair of increasing numbers such that w flips their order:

$$\text{inv}(w) = |\{(i, j) \mid i < j, w(i) > w(j)\}|.$$

Exercise 11.5 (a) Find $\text{inv}(w_0)$.
(b) Show that $\text{inv}(w) = \text{inv}(w^{-1})$.
(c) Show that $\text{inv}(ws_i) = \text{inv}(w) + 1$ if $w(i) < w(i+1)$ (i.e., if i is an *ascent* of the permutation w) and $\text{inv}(ws_i) = \text{inv}(w) - 1$ if $w(i) > w(i+1)$ (i.e., if i is a *descent* of w).

Let us prove the equivalence of the two definitions of length for a permutation.

Proposition 11.6 *Any shortest word for a permutation w consists of* $\text{inv}(w)$ *simple transpositions.*

Proof If $\text{inv}(w) = 0$, then w is the identity permutation, and the shortest word for it is empty.

Let us show that $\ell(w) \le \text{inv}(w)$ by induction over $\text{inv}(w)$. If w is not the identity, then it has a *descent* i, i.e., a number such that $w(i) > w(i+1)$. Then $\text{inv}(ws_i) = \text{inv}(w) - 1$. The induction hypothesis says that $\ell(ws_i) \le \text{inv}(w) - 1$. Let Q be a shortest word for ws_i. Then Qs_i is a word for w, hence $\ell(w) \le \ell(ws_i) + 1 \le \text{inv}(w)$.

Now let us prove the inverse inequality: $\text{inv}(w) \le \ell(w)$. Again, we proceed by induction over $\ell(w)$. Let Q be a reduced word for w. Deleting the last letter s_i in Q, we obtain a reduced word Q' of permutation w'. By the induction hypothesis we have $\text{inv}(w') \le \ell(w') = \ell(w) - 1$. Then $\text{inv}(w) = \text{inv}(w's_i) \le \text{inv}(w') + 1 \le \ell(w)$, as desired. □

Corollary 11.7 *A word is reduced if and only if any pair of strands in its wiring diagram intersects at most once.*

Proof If a certain pair of strands intersects twice, we can remove both these intersections. The permutation remains the same, and the length of the word decreases by 2, so the initial word was not reduced.

Now suppose that no pair of strands intersects twice. Then a pair (i, j) forms an inversion if and only if the strands ending at i and j intersect. Then the number of intersections (or, equivalently, of letters in our word) equals $\text{inv}(w)$, hence the word is reduced. □

11.3 The Graph of Reduced Words

Any pair of neighboring letters $s_i s_j$, where $|i - j| \geq 2$, in a reduced word for a permutation $w \in \mathcal{S}_n$ can be replaced by $s_j s_i$; this will produce another reduced word for the same permutation. Same thing can be done with a triple $s_i s_{i+1} s_i$, which can be replaced by $s_{i+1} s_i s_{i+1}$.

Let the reduced words for $w \in \mathcal{S}_n$ be the vertices of a graph and join by edges the words obtained one from another by such relations (i.e., by far commutativity and braid relations). We obtain the *graph of reduced words*. The next proposition states that this graph is always connected.

Proposition 11.8 *Any reduced word for a permutation w can be transformed into any other reduced word by applying far commutativity and braid relations.*

Proof Consider an arbitrary reduced word Q for the permutation w. We shall bring it to a certain canonical form. We will illustrate this with the wiring diagram of the following reduced word:

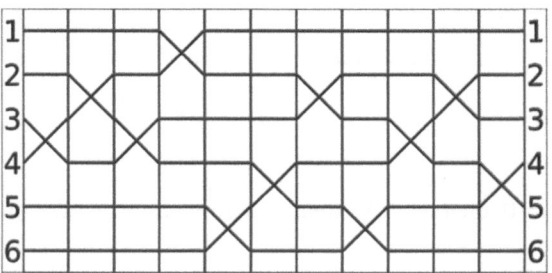

Let i be the smallest number satisfying $w(i) \neq i$ (in other words, it is the minimal i such that s_i occurs in Q). In our case $i = 1$. Consider the leftmost letter s_i in Q. Using far commutativity, shift it to the right until it reaches s_{i+1} (it cannot reach s_i, since the word is reduced, and the letter s_{i-1} does not occur in our word at all).

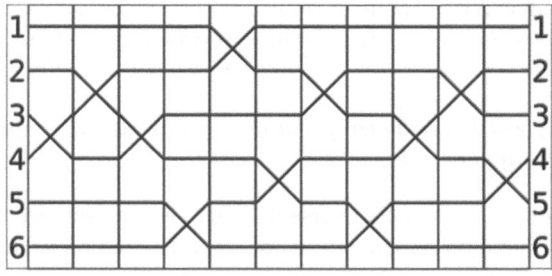

↓

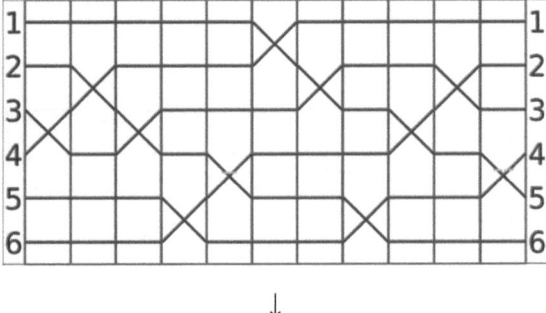

↓

Now let us shift the whole pair $s_i s_{i+1}$ to the right by far commutativity. If at some point it reaches s_{i+2}, we add this letter to the sequence that we are moving, and proceed further.

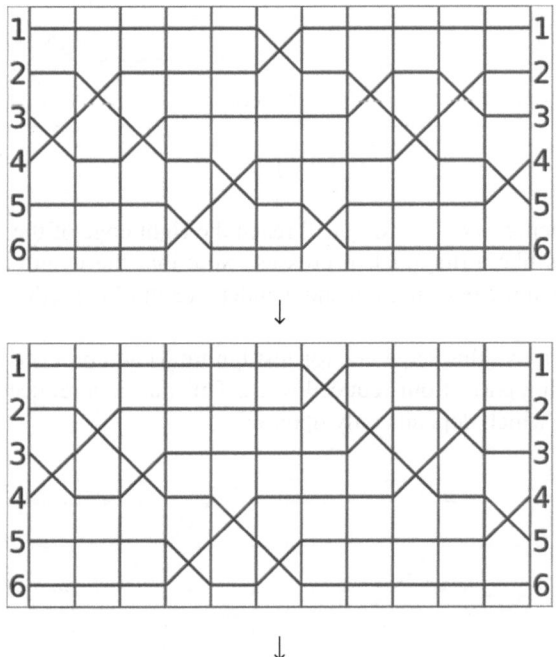

↓

Suppose that at some point the sequence $s_i \ldots s_{i+k}$ reaches s_{i+k-1}. Then apply the braid relation and move the whole sequence to the right:

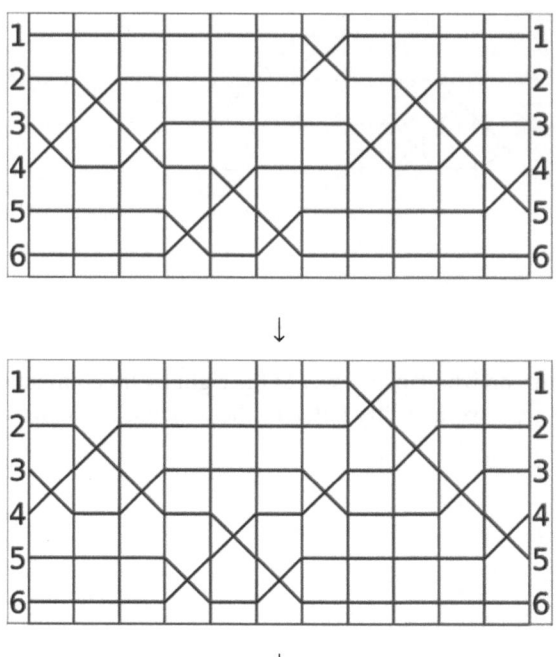

↓

↓

Finally this sequence $s_i s_{i+1} \ldots s_{i+k}$ will reach the right edge of the diagram. On the diagram it looks like a diagonal of crosses. Note that the length of this sequence depends only upon the permutation and is independent of the reduced word: indeed, $i + k = w^{-1}(i)$.

Now let us remove this sequence (or just ignore it) and proceed in a similar way with the remaining part of our reduced word. This part is a reduced word for some permutation w', which depends only upon w.

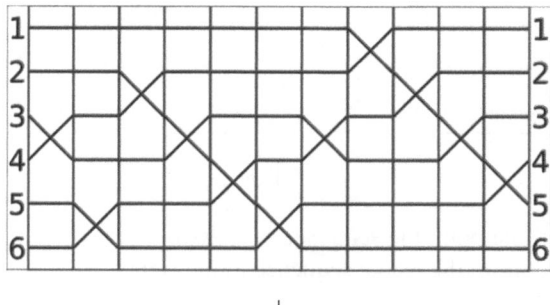

↓

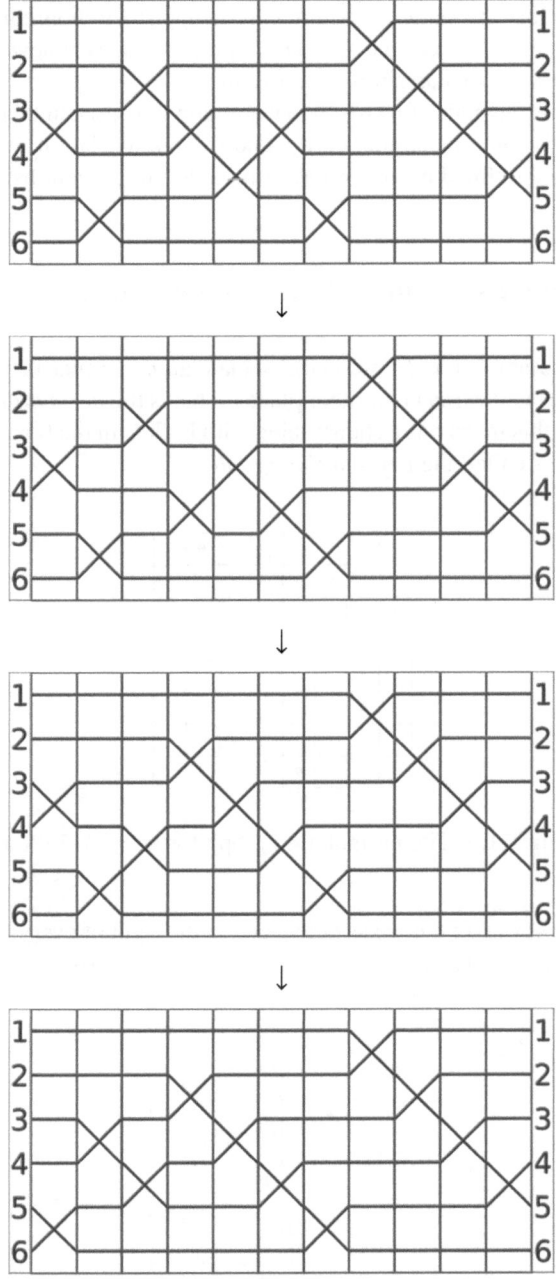

Finally we will get a diagram consisting of several diagonals. The exact form of this diagram depends only upon the permutation w. So every reduced word can be brought to such a form. This means that any pair of reduced words can be joined by a sequence of far commutativity and braid relations. □

Remark 11.9 Almost the same algorithm can be applied to nonreduced words. We start with the minimal i such that s_i occurs in our word. Whenever we obtain two consecutive letters s_k, we can simply remove them.

This means that any pair of words of permutation $w \in S_n$ can be joined by means of the relation $s_i^2 = e$, far commutativity, and the braid relation, implying that the group S_n is indeed defined by generators s_i modulo these relations.

11.4 The Rothe Diagram and the Lehmer Code

To each permutation $w \in S_n$ we can associate an $n \times n$-matrix W with 1s at the positions $(i, w(i))$ and zeros at all other places. This is the *permutation matrix* of w.

We will draw this matrix as a square table with bullets in the boxes $(i, w(i))$. This table is sometimes called the *permutation graph*.

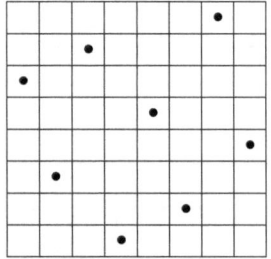

Fig. 11.2: The permutation graph for $w = \overline{73158264}$

Now let us cross out all the boxes below and to the right of every bullet (including the boxes with bullets). The remaining (uncrossed) boxes form the *Rothe diagram* of this permutation.

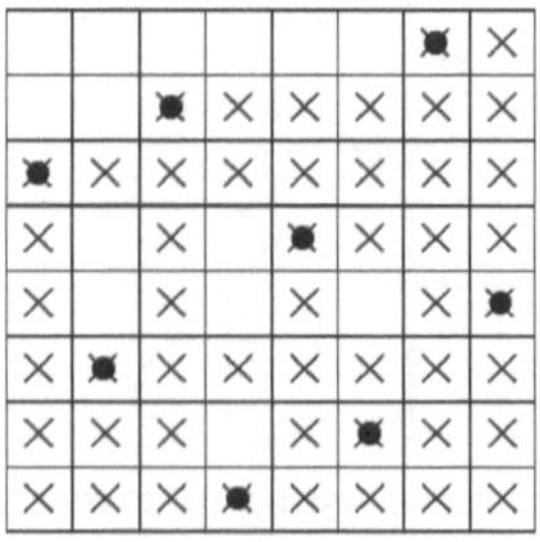

Fig. 11.3: The Rothe diagram for $w = \overline{73158264}$

Exercise 11.10 Prove the following:

(a) Every connected component of a Rothe diagram is a Young diagram.
(b) The number of boxes in the Rothe diagram for w equals inv(w).

Multiplication of w by s_i on the right corresponds to interchanging the rows i and $i + 1$.

Exercise 11.11 Show that this operation changes the number of boxes in the Rothe diagram by ± 1.

It is clear that the number of boxes in the i-th row of the Rothe diagram is equal to the number of $j > i$ such that $w(i) < w(j)$. This relates the Rothe diagram with the so-called Lehmer code of the permutation.

Definition 11.12 Given a permutation $w \in S_n$, the sequence

$$L(w) = (L_1(w), L_2(w), \ldots, L_n(w))$$

defined by $L_i(w) = |\{j > i \mid w(j) < w(i)\}|$ is called the *Lehmer code* of w.

Exercise 11.13

(a) Find the Lehmer code of $w = \overline{426351}$.
(b) Find a permutation with Lehmer code $(4, 0, 2, 0, 1, 1, 0)$.
(c) Show that a permutation is uniquely determined by its Lehmer code.

11.5 The Bruhat Order

Let us introduce a partial order on the elements of S_n. We shall write $w \lessdot w'$ if

$$w' = w \circ (i \leftrightarrow j), \qquad \ell(w') = \ell(w) + 1.$$

Remark 11.14 This definition can be reformulated in terms of Rothe diagrams. Consider the diagram for w with bullets in $(i, w(i))$ and $(j, w(j))$. To get the diagram for $w' = w \circ (i \leftrightarrow j)$, we need to interchange the rows containing these bullets; afterwards they will be situated in $(j, w(i))$ and $(i, w(j))$.

Since $\ell(w') = \ell(w) + 1$, the multiplication by a transposition should increase the number of uncrossed boxes in the diagram by 1. This means that they were initially situated in the upper-left and the bottom-right corners of a rectangle without any other bullets inside it.

Exercise 11.15 Prove the last statement.

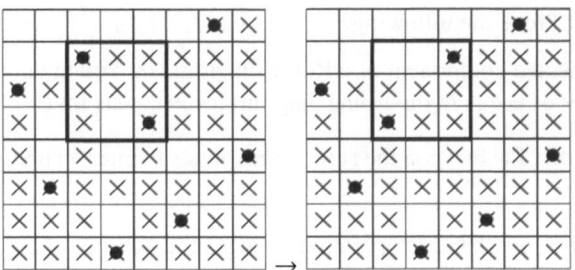

Fig. 11.4: $w = \overline{73158264} \prec \overline{75138264} = w \circ (2 \leftrightarrow 4)$

The *covering relation* $\prec$ can be extended by transitivity, providing a partial order $\leq$ on $\mathcal{S}_n$. It is called the *Bruhat order*.

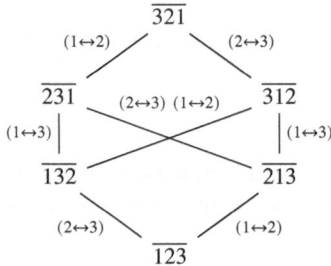

Fig. 11.5: The Bruhat order on $\mathcal{S}_3$

11.6 Problems

11.1 Write down a reduced word for the longest permutation

$$w_{0,n} = \overline{n, n-1, \ldots, 2, 1} \in \mathcal{S}_n.$$

11.2 Compute

$$\sum_{w \in \mathcal{S}_n} \ell(w).$$

11.3 Write down all reduced words for

(a) $w = \overline{4231} \in \mathcal{S}_4$
(b) $w = \overline{42153} \in \mathcal{S}_5$

and draw the corresponding graph of the reduced words.

11.4 A permutation $w \in S_n$ is said to be *k-Grassmannian* if it has a unique descent on the k-th position (i.e., $w(k) > w(k+1)$, while $w(i) < w(i+1)$ for all $i \neq k$). Let $L(w) = (L_1(w), L_2(w), \ldots, L_n(w))$ be its Lehmer code. Show that

$$L_k(w) \geq L_{k-1}(w) \geq \cdots \geq L_1(w)$$

and

$$L_{k+1}(w) = \cdots = L_n(w) = 0.$$

Construct a bijection between k-Grassmannian permutations and partitions of length k with parts not exceeding $n - k$.

11.5 a. Permutation $w \in S_n$ is said to be *dominant*, or *132-avoiding*, if there is no triple $i < j < k$ such that $w(i) < w(k) < w(j)$. What can be said about the Rothe diagram and the Lehmer code of such a permutation?
      b. Find the number of 132-avoiding permutations $w \in S_n$.

11.6 A permutation $w \in S_n$ is said to be *almost increasing*, or *321-avoiding*, if there is no triple $i < j < k$ such that $w(k) < w(j) < w(i)$.

   (a) Let $w \in S_n$ be a permutation. Denote by $i_1 < i_2 < \cdots < i_k$ the entries satisfying $w(i_s) \geq i_s$, and by $j_1 < j_2 < \cdots < j_{n-k}$ the entries satisfying $w(j_s) < j_s$. Show that w is 321-avoiding if and only if $w(i_1) < w(i_2) < \cdots < w(i_k)$ and $w(j_1) < w(j_2) < \cdots < w(j_{n-k})$.
   (b) Show that a permutation w is 321-avoiding if and only if any two reduced words for w can be obtained from each other by a sequence of far commutativity relations.
   (c) Show that the number of 321-avoiding permutations $w \in S_n$ equals the n-th Catalan number C_n.

11.7 A permutation $w \in S_n$ is called *vexillary*,[1] or *2143-avoiding*, if there is no quadruple $i < j < k < l$ such that $w(j) < w(i) < w(l) < w(k)$. Show that a permutation is vexillary if and only if its Rothe diagram can be transformed into a Young diagram by a suitable permutation of its rows and columns.

---

[1] This term comes from the word *vexillum*, the military standard of Roman legions. The only extant vexillum, dating back to the 3rd century CE, is now stored in the Pushkin Museum of Fine Arts in Moscow.

Chapter 12
Schubert Polynomials

This chapter is devoted to *partially symmetric* polynomials, i.e. polynomials that are symmetric with respect to *certain* changes of variables: Schubert polynomials. We will mostly follow A. Knutson's notes [Knu12].

12.1 Divided Difference Operators

Consider a polynomial ring $\mathbb{Z}[x_1, \ldots, x_n]$ with the action of the symmetric group $\mathcal{S}_n$ by changes of variables:

$$s_i \circ f(x_1, \ldots, x_i, x_{i+1}, \ldots, x_n) = f(x_1, \ldots, x_{i+1}, x_i, \ldots, x_n).$$

Note that the polynomial $f(\mathbf{x}) - s_i \circ f(\mathbf{x})$ is antisymmetric with respect to x_i and x_{i+1} and hence is divisible by $x_i - x_{i+1}$. This allows us to introduce divided difference operators.

Definition 12.1 A *divided difference operator* ∂_i, where $1 \le i \le n-1$, acts on the polynomial ring $\mathbb{Z}[x_1, \ldots, x_n]$ as follows

$$\partial_i f = \frac{f(\mathbf{x}) - s_i \circ f(\mathbf{x})}{x_i - x_{i+1}}.$$

Exercise 12.2

(a) Show that for a homogeneous polynomial f of degree k its image $\partial_i f$ is either a homogeneous polynomial of degree $k - 1$ or zero.
(b) Show that $\partial_i f = 0$ if and only if f is symmetric with respect to x_i and x_{i+1}.
(c) Let $f \in \mathbb{Z}[x_1, \ldots, x_n]$. Show that if we have $\partial_i f = 0$ for each $1 \le i \le n$, then $f = \text{const}$. (To apply ∂_n to f we consider f as an element of $\mathbb{Z}[x_1, \ldots, x_n, x_{n+1}]$.)

Obviously, the operators ∂_i are linear. We shall also need formulas describing the action of divided difference operators on products of polynomials, stated in the following exercise.

© The Author(s), under exclusive license to Springer Nature Switzerland AG 2024
E. Smirnov, A. Tutubalina, *Symmetric Functions: A Beginner's Course*,
Moscow Lectures 10, https://doi.org/10.1007/978-3-031-50341-2_12

Exercise 12.3

(a) If f is symmetric with respect to x_i and x_{i+1}, then $\partial_i(fg) = f\partial_i g$.

(b) In the general case the operators ∂_i satisfy the "twisted Leibniz rule":

$$\partial_i(fg) = (\partial_i f)g + (s_i \circ f)(\partial_i g).$$

This provides an analogy between ∂_i and partial derivatives. However, divided difference operators are more complicated: in particular, they do not necessarily commute. The commutation rules for them resemble those for simple transpositions; namely, we have:

- $\partial_i^2 = 0$;
- $\partial_i \partial_j = \partial_j \partial_i$ for $|i - j| \geq 2$;
- $\partial_i \partial_{i+1} \partial_i = \partial_{i+1} \partial_i \partial_{i+1}$.

Exercise 12.4 Prove these relations.

The second and the third relations are identical to the far commutativity and braid relations for s_i. However, the relation $\partial_i^2 = 0$ is different from $s_i^2 = e$. The operators ∂_i form a so-called *nil-Hecke algebra*; it can be obtained as a deformation from the group algebra of the symmetric group.

These relations also allow us to define divided difference operators for arbitrary permutations $w \in S_n$.

Definition 12.5 Let $w \in S_n$ be a permutation, and let $s_{i_1} s_{i_2} \ldots s_{i_k}$ be a reduced word for w. Then ∂_w is defined as

$$\partial_w = \partial_{i_1} \partial_{i_2} \ldots \partial_{i_k}.$$

This operator is well-defined, i.e., independent of a choice of reduced word. Indeed, Proposition 11.8 guarantees that any two reduced words can be obtained one from another by means of far commutativity and braid relations. But the operators ∂_i satisfy these relations as well.

Exercise 12.6 Let $u, v \in S_n$, $w = uv$. Show that

$$\partial_u \partial_v = \begin{cases} \partial_w, & \text{if } \ell(u) + \ell(v) = \ell(w), \\ 0, & \text{if } \ell(u) + \ell(v) > \ell(w). \end{cases}$$

12.2 Schubert Polynomials

Definition 12.7 *Schubert polynomials*[1] are defined as a family of homogeneous polynomials

$$\mathfrak{S}_w \in \mathbb{Z}[x_1, \ldots, x_{n-1}]$$

---

[1] Schubert polynomials were first defined by A. Lascoux and M.-P. Schützenberger in [LS82].

indexed by $w \in S_n$ and satisfying the following recurrence relations:

$$\mathfrak{S}_e = 1,$$

$$\partial_i \mathfrak{S}_w = \begin{cases} \mathfrak{S}_{ws_i}, & \text{if } \ell(ws_i) < \ell(w), \\ 0, & \text{if } \ell(ws_i) > \ell(w), \end{cases}$$

for each $i = 1, \ldots, n-1$.

Remark 12.8 Note that the condition $\ell(ws_i) < \ell(w)$ is equivalent to $w(i) > w(i+1)$, i.e. to the fact that i is a descent for the permutation w.

This definition has an obvious drawback: it neither guarantees the existence of such polynomials nor their uniqueness; we will need to prove them. Let us start with uniqueness.

Proposition 12.9 *There exists at most one family of polynomials satisfying Definition 12.7.*

Proof Suppose that we have two such families, $\mathfrak{S}_w$ and $\widetilde{\mathfrak{S}}_w$. We shall proceed by induction over $\ell(w)$. The induction base is obvious: $\mathfrak{S}_e = \widetilde{\mathfrak{S}}_e = 1$.

Let w be a shortest permutation satisfying $\mathfrak{S}_w \neq \widetilde{\mathfrak{S}}_w$. Then for every i such that $\ell(ws_i) < \ell(w)$ we have

$$\partial_i \left(\mathfrak{S}_w - \widetilde{\mathfrak{S}}_w \right) = \mathfrak{S}_{ws_i} - \widetilde{\mathfrak{S}}_{ws_i} = 0.$$

Clearly, for all other i's we have $\ell(ws_i) > \ell(w)$, so the equality $\partial_i(\mathfrak{S}_w) - \partial_i(\widetilde{\mathfrak{S}}_w) = 0$ also holds, since both terms in the left-hand side are zero.

Then Exercise 12.2 implies that $\mathfrak{S}_w - \widetilde{\mathfrak{S}}_w = \text{const}$, and this constant is nonzero. But both these polynomials are homogeneous of degree $\ell(w)$ (indeed, every ∂_i decreases the degree of the polynomial by 1, so to obtain $\mathfrak{S}_e$ from $\mathfrak{S}_w$ we need to apply ∂_i exactly $\ell(w)$ times). Contradiction. $\qquad\qquad\square$

Now let us present an explicit construction for Schubert polynomials. It will prove their existence and imply that Definition 12.7 indeed makes sense.

Proposition 12.10 *Let* $w_{0,n} = \overline{n, n-1, \ldots, 2, 1} \in S_n$ *be the longest permutation. Then*

$$\mathfrak{S}_{w_{0,n}} = x_1^{n-1} x_2^{n-2} \ldots x_{n-1}.$$

The Schubert polynomials of other permutations are obtained as follows:

$$\mathfrak{S}_w = \partial_{w^{-1} w_{0,n}} \mathfrak{S}_{w_{0,n}}.$$

Exercise 12.11 Show that $\ell(w^{-1} w_{0,n}) = \ell(w_{0,n}) - \ell(w)$.

Proof First let us check the recurrence relations:

$$\partial_i \mathfrak{S}_w = \partial_i \partial_{w^{-1}w_{0,n}} \mathfrak{S}_{w_{0,n}}$$

$$= \begin{cases} \partial_{s_i w^{-1}w_{0,n}} \mathfrak{S}_{w_{0,n}} = \partial_{(ws_i)^{-1}w_{0,n}} \mathfrak{S}_{w_{0,n}} \\ \qquad\qquad = \mathfrak{S}_{ws_i}, & \text{if } \ell\left(s_i w^{-1}w_{0,n}\right) > \ell(w^{-1}w_{0,n}) + 1 \\ 0, & \text{otherwise.} \end{cases}$$

The first case occurs if and only if $\ell\left(s_i w^{-1}w_{0,n}\right) = \ell(w^{-1}w_{0,n}) + 1$, and this is equivalent to the desired equality $\ell(ws_i) = \ell(w) - 1$.

Now check the initial condition. For this, fix the following reduced word for $w_{0,n}$:

$$w_{0,n} = (s_1)\,(s_2 s_1)\,(s_3 s_2 s_1)\ldots(s_{n-1}s_{n-2}\ldots s_2 s_1).$$

Since

$$\partial_{n-1}\partial_{n-2}\ldots\partial_3\partial_2\partial_1 \left(x_1^{n-1}x_2^{n-2}x_3^{n-3}\ldots x_{n-2}^2 x_{n-1}\right)$$

$$= \partial_{n-1}\partial_{n-2}\ldots\partial_3\partial_2 \left(x_1^{n-2}x_2^{n-2}x_3^{n-3}\ldots x_{n-2}^2 x_{n-1}\right)$$

$$= \partial_{n-1}\partial_{n-2}\ldots\partial_3 \left(x_1^{n-2}x_2^{n-3}x_3^{n-3}\ldots x_{n-2}^2 x_{n-1}\right) = \ldots$$

$$= x_1^{n-2}x_2^{n-3}x_3^{n-4}\ldots x_{n-2},$$

we have $\partial_{w_{0,n}} \mathfrak{S}_{w_{0,n}} = 1$, as desired. $\qquad\qquad\qquad\qquad\qquad\qquad\square$

Exercise 12.12 In this construction Schubert polynomials may *a priori* depend upon n. Show that this is not true: if we consider a permutation from S_n as a permutation from S_{n+1} leaving $n + 1$ fixed, this does not change the Schubert polynomial $\mathfrak{S}_w$. We shall say that Schubert polynomials are *stable* with respect to the embedding $S_n \hookrightarrow S_{n+1}$.

Exercise 12.13

(a) Show that $\mathfrak{S}_{s_i} = x_1 + \cdots + x_i$.
(b) Compute $\mathfrak{S}_w$ for all $w \in S_3$.

12.3 Monk's Rule

In this section we deduce an analogue of the Pieri formulas for Schubert polynomials. Namely, we express the product $\mathfrak{S}_{s_i}\mathfrak{S}_w$ as a linear combination of Schubert polynomials.

Theorem 12.14 (Monk's rule) *For a permutation $w \in S_n$ and a simple transposition $s_i \in S_n$ we have:*

$$\mathfrak{S}_{s_i}\mathfrak{S}_w = \sum_{\substack{k \leq i < m \\ \ell(w\circ(k\leftrightarrow m))=\ell(w)+1}} \mathfrak{S}_{w\circ(k\leftrightarrow m)}. \qquad\qquad (12.1)$$

Proof Both sides of this equality are homogeneous polynomials of degree $\ell(w) + 1$. Then it is enough to show that

$$\partial_j \left(\mathfrak{S}_{s_i} \mathfrak{S}_w \right) = \partial_j \left(\sum_{\substack{k \leq i < m \\ \ell(w \circ (k \leftrightarrow m)) = \ell(w) + 1}} \mathfrak{S}_{w \circ (k \leftrightarrow m)} \right) \tag{12.2}$$

for each j ranging from 1 to $n - 1$.

We argue by induction on $\ell(w)$. Suppose Monk's rule holds for all permutations v satisfying $\ell(v) < \ell(w)$.

In the remaining part of the proof we need to consider several cases; namely, j can be either a descent or an ascent in w, and i may or may not be equal to j.

In general, this proof is simple, but long and laborious. We recommend the readers to work through it by themselves, using, where necessary, graphs of permutations.

- Let $j \neq i$ and let j be an ascent in w (i.e., $w(j) < w(j+1)$). Then both $\mathfrak{S}_{s_i}$ and $\mathfrak{S}_w$ are symmetric with respect to x_j and x_{j+1}, so the left-hand side in (12.2) is zero.

  Now consider a summand $\mathfrak{S}_{w \circ (k \leftrightarrow m)}$ from the left-hand side. As shown in Remark 11.14, the points $(k, w(k))$ and $(m, w(m))$ of the permutation graph of w are the top left and bottom right corner of a rectangle **r** that does not contain any other points of the graph inside it.

  - If $k = j$, we have $w(k) < w(j+1)$. Moreover, if $k \neq i$, then $k < i < m$, which implies $m \neq j + 1$. This means that the point $(j + 1, w(j + 1))$ is located to the right of the rectangle r, which means that $w(m) < w(j+1)$. So the permutation graph has the following form:

  Now draw the graph of the permutation $w \circ (k \leftrightarrow m)$:

We see that j is an ascent for this permutation as well. Then $\partial_j \mathfrak{S}_{w \circ (k \leftrightarrow m)} = 0$.

- If $k = j + 1$, we have $w(j) < w(j + 1) = w(k) < w(m)$.

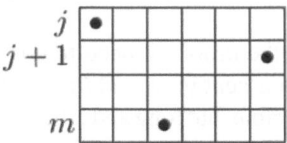

So j is an ascent in $w \circ (k \leftrightarrow m)$, and $\partial_j \mathfrak{S}_{w \circ (k \leftrightarrow m)} = 0$.

- The cases $m = j$ and $m = j + 1$ are treated similarly. If both k and m are distinct from j and $j + 1$, the permutations s_j and $(k \leftrightarrow m)$ commute, so the number j still remains an ascent for the permutation $w \circ (k \leftrightarrow m)$.

This shows that the right-hand side of (12.2) also equals 0.

- Let $j = i$ and let j be an ascent in w. Then $\mathfrak{S}_w$ is symmetric with respect to x_j and x_{j+1}, and we have

$$\partial_i \left(\mathfrak{S}_{s_i} \mathfrak{S}_w \right) = \mathfrak{S}_w \partial_i \mathfrak{S}_{s_i} = \mathfrak{S}_w.$$

Now consider the right-hand side of (12.2). The number i remains an ascent for all permutations $w \circ (k \leftrightarrow m)$ (the proof is similar to the previous case), except the situation when $k = i$ and $m = i + 1$. Such a summand indeed exists, since $\ell(w s_i) > \ell(w)$.

Summarizing, we see that all the summands in the right-hand side of (12.2) are zero, except $\partial_i \mathfrak{S}_{w s_i} = \mathfrak{S}_w$. This equals the left-hand side, as desired.

- Now let $j \neq i$ and let j be a descent in w (i.e., $w(j) > w(j + 1)$). Then $\mathfrak{S}_{s_i}$ is symmetric with respect to x_j and x_{j+1}, and hence

$$\partial_j \left(\mathfrak{S}_{s_i} \mathfrak{S}_w \right) = \mathfrak{S}_{s_i} \partial_j \mathfrak{S}_w = \mathfrak{S}_{s_i} \mathfrak{S}_{w s_j}.$$

Since $\ell(w s_j) < \ell(w)$, we can use the induction hypothesis and expand this product using Monk's rule:

$$\mathfrak{S}_{s_i} \mathfrak{S}_{w s_j} = \sum_{\substack{k \le i < m \\ \ell(w s_j \circ (k \leftrightarrow m)) = \ell(w s_j) + 1}} \mathfrak{S}_{w s_j \circ (k \leftrightarrow m)}. \tag{12.3}$$

The right-hand side of (12.2) can be written as follows:

$$\sum_{\substack{k \le i < m \\ \ell(w \circ (k \leftrightarrow m)) = \ell(w) + 1 \\ j \text{ descent in } w \circ (k \leftrightarrow m)}} \mathfrak{S}_{w \circ (k \leftrightarrow m) s_j}. \tag{12.4}$$

We need to show that the expressions (12.3) and (12.4) consist of the same summands. Again for this we will use the permutation graph for w and consider all "minimal" rectangles with corners in $(k, w(k))$ and $(m, w(m))$, from (12.1). By minimality we mean that these rectangles do not contain other graph points inside them.

– Consider pairs of rectangles with upper left corners $(j, w(j))$ and $(j + 1, w(j + 1))$ and bottom right corner $(m, w(m))$. Here $w(j + 1) < w(j) < w(m)$. (The inequalities $i \geq k$ and $i \neq j$ imply that $i \geq j + 1$.)

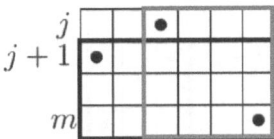

Now consider the graph of the permutation $w s_j$:

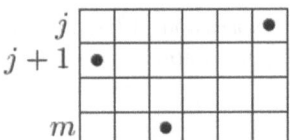

In the formula (12.3) only one of these two rectangles survives: the one with corners in $(j + 1, w(j))$ and $(m, w(m))$ (since $i \geq j + 1$, it does not contain a rectangle with corners $(j, w(j + 1))$ and $(j + 1, w(j))$). Thus, from the two summands of the initial formula in (12.3) we are left only with the Schubert polynomial of the permutation

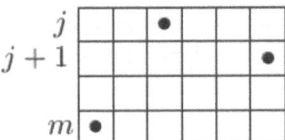

In the right-hand side of (12.1) we have two Schubert polynomials corresponding to the permutations

The number j is a descent only for the first one, so in (12.4) we are left only with the Schubert polynomial of the permutation

So these summands in (12.3) and (12.4) coincide.

– The permutation graph of w may also contain rectangles with corners $(j + 1, w(j + 1))$ and $(m, w(m))$, where $w(j + 1) < w(m) < w(j)$.

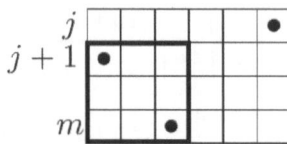

The permutation graph for ws_j looks as follows:

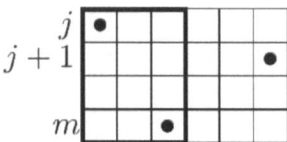

In the expression (12.3) this rectangle corresponds to the Schubert polynomial of the permutation

The right-hand side of (12.1) contains the Schubert polynomial of the permutation

The number j is a descent for it, so the expression (12.4) also contains the Schubert polynomial corresponding to the permutation

which coincides with the polynomial from (12.3).

- The cases $m = j$ and $m = j + 1$ are treated similarly. And if both k and m are distinct from j and $j + 1$, then the transpositions s_j and $(k \leftrightarrow m)$ commute, so the corresponding summands in (12.3) and (12.4) also coincide.

- The last remaining case corresponds to $j = i$, with i being a descent in w. Compute the left-hand side in (12.2). We shall use the "twisted Leibniz rule" for divided difference operators:

$$
\begin{aligned}
\partial_i \left(\mathfrak{S}_{s_i} \mathfrak{S}_w \right) &= \mathfrak{S}_w \left(\partial_i \mathfrak{S}_{s_i} \right) + \left(s_i \circ \mathfrak{S}_{s_i} \right) \partial_i \mathfrak{S}_w \\
&= \mathfrak{S}_w + (x_1 + \cdots + x_{i-1} + x_{i+1}) \mathfrak{S}_{w s_i} \\
&= \mathfrak{S}_w + \left(\mathfrak{S}_{s_{i+1}} - \mathfrak{S}_{s_i} + \mathfrak{S}_{s_{i-1}} \right) \mathfrak{S}_{w s_i}.
\end{aligned}
$$

Since $\ell(w s_i) < \ell(w)$, we can use the induction hypothesis to rewrite this expression using Monk's rule:

$$
\partial_i \left(\mathfrak{S}_{s_i} \mathfrak{S}_w \right) = \mathfrak{S}_w + \left(\sum_{\substack{k \le i+1 < m \\ \ell(w s_i \circ (k \leftrightarrow m)) = \ell(w s_i)+1}} \mathfrak{S}_{w s_i \circ (k \leftrightarrow m)} \right)
$$
$$
- \left(\sum_{\substack{k \le i < m \\ \ell(w s_i \circ (k \leftrightarrow m)) = \ell(w s_i)+1}} \mathfrak{S}_{w s_i \circ (k \leftrightarrow m)} \right)
$$
$$
+ \left(\sum_{\substack{k \le i-1 < m \\ \ell(w s_i \circ (k \leftrightarrow m)) = \ell(w s_i)+1}} \mathfrak{S}_{w s_i \circ (k \leftrightarrow m)} \right).
$$

Each of the summands can occur in one, two, or three brackets. The following table indicates the coefficient in front of each summand in the resulting sum.

	$k \le i - 1$	$k = i$	$k = i + 1$
$m = i$	1	0	0
$m = i + 1$	0	−1	0
$m > i + 1$	1	0	1

Thus, we have

$$
\partial_i \left(\mathfrak{S}_{s_i} \mathfrak{S}_w \right) = \mathfrak{S}_w + \left(\sum_{\substack{k \le i-1 \\ \ell(w s_i \circ (k \leftrightarrow i)) = \ell(w s_i)+1}} \mathfrak{S}_{w s_i \circ (k \leftrightarrow i)} \right) - \mathfrak{S}_{w s_i \circ (i \leftrightarrow i+1)}
$$

$$
+ \left(\sum_{\substack{k \le i-1 \\ m > i+1 \\ \ell(w s_i \circ (k \leftrightarrow m)) = \ell(w s_i)+1}} \mathfrak{S}_{w s_i \circ (k \leftrightarrow m)} \right) + \left(\sum_{\substack{i+1 < m \\ \ell(w s_i \circ (i+1 \leftrightarrow m)) = \ell(w s_i)+1}} \mathfrak{S}_{w s_i \circ (i+1 \leftrightarrow m)} \right)
$$

$$
= \left(\sum_{\substack{k \le i-1 \\ \ell(ws_i \circ (k \leftrightarrow i)) = \ell(ws_i)+1}} \mathfrak{S}_{wo(k \leftrightarrow i+1) \circ s_i} \right) + \left(\sum_{\substack{k \le i-1 \\ m > i+1 \\ \ell(wo(k \leftrightarrow m) \circ s_i) = \ell(ws_i)+1}} \mathfrak{S}_{wo(k \leftrightarrow m) \circ s_i} \right)
$$

$$
+ \left(\sum_{\substack{i+1 < m \\ \ell(ws_i \circ (i+1 \leftrightarrow m)) = \ell(ws_i)+1}} \mathfrak{S}_{wo(i \leftrightarrow m) \circ s_i} \right). \quad (12.5)
$$

Now look at the right-hand side of (12.2). Since i is a descent, this sum does not contain a summand for $k = i$ and $m = i+1$. The remaining summands can be split into three parts:

$$
\partial_i \left(\sum_{\substack{k \le i < m \\ \ell(wo(k \leftrightarrow m)) = \ell(w)+1}} \mathfrak{S}_{wo(k \leftrightarrow m)} \right) =
$$

$$
\left(\sum_{\substack{k \le i-1 \\ \ell(wo(k \leftrightarrow i+1)) = \ell(w)+1 \\ i \text{ descent for } wo(k \leftrightarrow i+1)}} \mathfrak{S}_{wo(k \leftrightarrow i+1) \circ s_i} \right) + \left(\sum_{\substack{k \le i-1 \\ m > i+1 \\ \ell(wo(k \leftrightarrow m)) = \ell(w)+1 \\ i \text{ descent for } wo(k \leftrightarrow m)}} \mathfrak{S}_{wo(k \leftrightarrow m) \circ s_i} \right)
$$

$$
+ \left(\sum_{\substack{m > i+1 \\ \ell(wo(i \leftrightarrow m)) = \ell(w)+1 \\ i \text{ descent for } wo(i \leftrightarrow m)}} \mathfrak{S}_{wo(i \leftrightarrow m) \circ s_i} \right). \quad (12.6)
$$

Finally, note that the corresponding brackets in (12.5) and (12.6) coincide. Indeed, the summation conditions for the first brackets in both (12.5) and (12.6) are equivalent to the following form of the permutation graph for w:

The transpositions s_i and $(k \leftrightarrow m)$ commute, which implies that the second brackets in both expressions are equal. The equality of the third brackets is obtained similarly to the first case.

This completes the proof.   □

12.4 Lascoux's Transition Formula

We can use the relation $x_i = \mathfrak{S}_{s_i} - \mathfrak{S}_{s_{i-1}}$ and Monk's rule to obtain a formula for multiplying Schubert polynomials by a variable x_i.

Corollary 12.15 *We have the following rule for multiplying Schubert polynomials by a variable:*

$$x_i \mathfrak{S}_w = \sum_{\substack{k \le i < m \\ \ell(w\circ(k\leftrightarrow m))=\ell(w)+1}} \mathfrak{S}_{w\circ(k\leftrightarrow m)} - \sum_{\substack{k \le i-1 < m \\ \ell(w\circ(k\leftrightarrow m))=\ell(w)+1}} \mathfrak{S}_{w\circ(k\leftrightarrow m)}$$

$$= \sum_{\substack{m > i \\ \ell(w\circ(i\leftrightarrow m))=\ell(w)+1}} \mathfrak{S}_{w\circ(i\leftrightarrow m)} - \sum_{\substack{k < i \\ \ell(w\circ(k\leftrightarrow i))=\ell(w)+1}} \mathfrak{S}_{w\circ(k\leftrightarrow i)}.$$

This corollary looks somewhat simpler than Monk's rule. But it has an obvious drawback: it contains minus signs, so it is not *manifestly positive*. It turns out that for some carefully chosen i it may become positive. For this, collect the negative summands in the left-hand side and the positive ones in the right-hand side and then pick i in such a way that the right-hand side contains only one summand.

Theorem 12.16 (Lascoux's transition formula) *Let $w \ne e$. Suppose that i is the last descent in w. Let j be the biggest number such that $w(j) < w(i)$ (obviously, $j \ge i+1$). Denote the permutation $w \circ (i \leftrightarrow j)$ by w'. Then we have*

$$\mathfrak{S}_w = x_i \mathfrak{S}_{w'} + \sum_{\substack{k < i \\ \ell(w'\circ(k\leftrightarrow i))=\ell(w')+1}} \mathfrak{S}_{w'\circ(k\leftrightarrow i)}.$$

Proof Applying Corollary 12.15 to the product $x_i \mathfrak{S}_{w'}$, we obtain that

$$x_i \mathfrak{S}_{w'} = \sum_{\substack{m > i \\ \ell(w'\circ(i\leftrightarrow m))=\ell(w')+1}} \mathfrak{S}_{w'\circ(i\leftrightarrow m)} - \sum_{\substack{k < i \\ \ell(w'\circ(k\leftrightarrow i))=\ell(w')+1}} \mathfrak{S}_{w'\circ(k\leftrightarrow i)}.$$

Now note that the first sum in the right-hand side contains only one summand, corresponding to $m = j$. Indeed, if $m < j$, we have $w(m) < w(j)$ (since i is the last descent), and thus $w'(m) < w'(i)$.

If $m > j$, we have $w(m) > w(i)$ (according to the choice of j). Then we have the inequalities $i < j < m$ and $w'(i) < w'(j) < w'(m)$, so the rectangle on the permutation graph of w with corners in $(i, w'(i))$ and $(m, w'(m))$ contains one more point (i.e., is not minimal).

Thus the first sum consists of a unique summand $\mathfrak{S}_{w'\circ(i\leftrightarrow j)} = \mathfrak{S}_w$. So we have

$$x_i \mathfrak{S}_{w'} = \mathfrak{S}_w - \sum_{\substack{k < i \\ \ell(w'\circ(k\leftrightarrow i))=\ell(w')+1}} \mathfrak{S}_{w'\circ(k\leftrightarrow i)},$$

which implies Lascoux's transition formula.   □

Introduce a linear order on permutations in the following way: order them by length, and order permutations of the same length reverse-lexicographically (according to their one-line notation). Then Lascoux's transition formula allows us to express Schubert polynomials of larger permutations via those of smaller ones with positive coefficients. Indeed, $\ell(w') = \ell(w) - 1$, and $\ell(w' \circ (k \leftrightarrow i)) = \ell(w)$, but the permutation $w = w' \circ (i \leftrightarrow j)$ is lexicographically smaller than $w' \circ (k \leftrightarrow i)$. This implies the following result.

Corollary 12.17 *All coefficients of Schubert polynomials are nonnegative.*

12.5 Problems

12.1 (a) Let $v, w \in S_n$, $\ell(v) = \ell(w)$. Show that $\partial_v \mathfrak{S}_w = \delta_{vw}$.
   (b) Using the previous exercise, show that Schubert polynomials are linearly independent.

12.2 Compute the Schubert polynomial $\mathfrak{S}_w$ for:

   (a) $w = s_i s_j$, where $|i - j| \geq 2$;
   (b) $w = \overline{1432}$.

12.3 Let $w \in S_n$ be a k-Grassmannian permutation. Show that the Schubert polynomial $\mathfrak{S}_w$ depends only upon the first k variables $x_1, \ldots, x_k$ and is symmetric with respect to these variables.

12.4 Let $k < n$. Find permutations $w \in S_n$ corresponding to the following Schubert polynomials:

   (a) $\mathfrak{S}_w = x_1 x_2 \ldots x_{n-1}$;
   (b) $\mathfrak{S}_w = x_1^{n-1}$;
   (c) $\mathfrak{S}_w = e_k (x_1, \ldots, x_{n-1})$;
   (d) $\mathfrak{S}_w = h_k (x_1, \ldots, x_{n-k})$.

12.5 Using Lascoux's transition formula, compute the Schubert polynomials for all $w \in S_4$.

Chapter 13
Combinatorial Presentation of Schubert Polynomials

In the previous chapter we defined Schubert polynomials by means of divided difference operators. Despite the fact that the differences are *divided*, Schubert polynomials are actual polynomials, not rational functions. More surprisingly, even though these are divided *difference* operators, the coefficients of Schubert polynomials turn out to be *positive*, as we saw in Corollary 12.17.

This reminds us of the situation with Schur polynomials. We defined them as ratios of two determinants (i.e., polynomials with many minus signs), but they turned out to be polynomials with nonnegative integer coefficients. This follows from Littlewood's theorem, providing a *combinatorial definition* of Schur polynomials: each of its coefficients is equal to the number of Young tableaux of given shape and weight.

It turns out that a similar combinatorial formula exists for Schubert polynomials (and, as we will see, it generalizes Littlewood's theorem). This formula was obtained by S. Billey and N. Bergeron in [BB93] and independently by S. Fomin and An. Kirillov in [FK96].

13.1 Pipe Dreams

Definition 13.1 Consider an $n \times n$ square and fill it with two kinds of elements: *crosses* $+$ and *elbow joints*, or just *elbows*: $\overset{\scriptstyle\frown}{r}$ in such a way that all crosses are located strictly above the antidiagonal. Such an object is called a *pipe dream*.[1]

Each pipe dream can be viewed as a configuration of strands (sometimes we will refer to them as to *pipes*, following the "plumber's terminology"). We will say that these strands enter from the left edge of the square and go to the top one. Index the endpoints of the strands by the integers from 1 to n from left to right and from top to bottom.

---

[1] This term, proposed by Allen Knutson, refers to the video game *Pipe Dream* by LucasArts, Inc., where the player constructs a system of pipelines from random elements.

© The Author(s), under exclusive license to Springer Nature Switzerland AG 2024
E. Smirnov, A. Tutubalina, *Symmetric Functions: A Beginner's Course*,
Moscow Lectures 10, https://doi.org/10.1007/978-3-031-50341-2_13

A pipe dream is said to be *reduced* if any pair of its strands intersects at most once. We will further speak only about reduced pipe dreams, so sometimes the word "reduced" will be omitted.

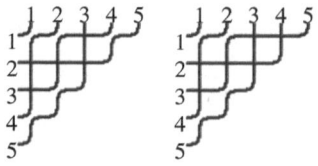

Fig. 13.1: A reduced (left) and a non-reduced (right) pipe dream

Definition 13.2 Each pipe dream defines a bijection between the sets of left and top endpoints of strands. For a pipe dream P, this bijection defines a permutation $w(P) \in S_n$; we will call it the *shape* of P. The set of all reduced pipe dreams of a given shape $w \in S_n$ will be denoted by $\mathrm{PD}(w)$.

For example, the shape of the left pipe dream P in Fig. 13.1 equals $w(P) = \overline{15423}$.

Exercise 13.3 Draw all pipe dreams of shape $\overline{1432}$.

A (not necessarily reduced) pipe dream P also defines a word Q_P. For this, let us "read" the pipe dream row by row, from top to bottom, reading each row *from right to left*. For each cross in row i and column j we will write the letter s_{i+j-1}; the elbows will simply be omitted. For instance, the word defined by the left pipe dream P in Fig. 13.1 equals $Q_P = s_3 s_4 s_3 s_2 s_3$.

Exercise 13.4 Using wiring diagrams, prove the following statements:

(a) Q_P is a word for the permutation $w(P)$.
(b) The word Q_P is reduced if and only if its pipe dream P is reduced.

This exercise implies that each reduced pipe dream of shape w contains exactly $\ell(w)$ crosses.

Consider a reduced pipe dream. Take two strands that intersect ($+$) in a box $\boxed{c}$ and "nearly meet" ($\curvearrowright$) in some other box $\boxed{e}$. Then we can move this cross from $\boxed{c}$ to $\boxed{e}$; clearly, we obtain a pipe dream of the same shape.

13.2 The Bergeron–Billey–Fomin–Kirillov Theorem

Schur polynomials are sums of monomials corresponding to Young tableaux of a given shape. It turns out that Schubert polynomials admit a similar description: they are sums of monomials corresponding to pipe dreams of a given shape

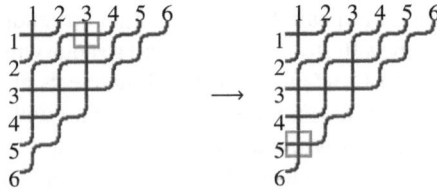

Fig. 13.2: Moving a cross in a pipe dream of shape $\overline{216543}$

We first describe the way to assign a monomial to a pipe dream. To each cross in the i-th row of a pipe dream we assign the weight x_i. The *weight* $\mathbf{x}^P$ of a pipe dream P is defined as the product of weights for all its crosses; i.e., the degree of occurrence of x_i in $\mathbf{x}^P$ equals the number of crosses in the i-th row of P.

Theorem 13.5 (N. Bergeron, S. Billey; S. Fomin, An. Kirillov) *The Schubert polynomial $\mathfrak{S}_w$ is equal to the sum of monomials $\mathbf{x}^P$ over all reduced pipe dreams of shape $w \in S_n$:*

$$\mathfrak{S}_w = \sum_{P \in \mathrm{PD}(w)} \mathbf{x}^P. \tag{13.1}$$

Exercise 13.6 Check this theorem for all $w \in S_3$ and for $\overline{1432} \in S_4$.

Proof Let us temporarily denote the "combinatorially defined Schubert polynomials", i.e. the right-hand sides in 13.1, by $\mathfrak{P}_w$, and check that they satisfy the Lascoux transition formula:

$$\mathfrak{P}_w = x_i \mathfrak{P}_{w'} + \sum_{\substack{k < i \\ \ell(w' \circ (k \leftrightarrow i)) = \ell(w') + 1}} \mathfrak{P}_{w' \circ (k \leftrightarrow i)},$$

where i is the last descent for the permutation w, and j is the maximal number satisfying $w(j) < w(i)$ and $w' = w \circ (i \leftrightarrow j)$.

Since $\mathfrak{P}_e = \mathfrak{S}_e = 1$, by induction this would imply the equality 13.1 for each $w \in S_n$. To check that the polynomials $\mathfrak{P}_w$ satisfy the Lascoux transition formula, construct a bijection

$$f : \mathrm{PD}(w) \to \mathrm{PD}(w') \cup \bigcup_{\substack{k < i \\ \ell(w' \circ (k \leftrightarrow i)) = \ell(w') + 1}} \mathrm{PD}\,(w' \circ (k \leftrightarrow i))$$

satisfying the condition

$$\mathbf{x}^P = x_i \cdot \mathbf{x}^{f(P)}, \text{ if } f(P) \in \mathrm{PD}(w')$$

$$\mathbf{x}^P = \mathbf{x}^{f(P)} \text{ otherwise.}$$

By the *strand number* we mean the number of its left endpoint. Note that in any pipe dream $P \in PD(w)$ the i-th strand intersects all strands with numbers from $i+1$ to j. In the pipe dream $P' \in PD(w')$ the i-th strand intersects all strands with numbers between $i + 1$ and $j - 1$ and does not intersect the strand with number j.

Also note that if the strands with numbers k and i do not intersect in a given pipe dream $P' \in PD(w')$, but "nearly meet" (form an elbow) at some box, then replacing this elbow by a cross turns P' into a reduced pipe dream of shape $w' \circ (k \leftrightarrow i)$. Moreover, $\ell(w' \circ (k \leftrightarrow i)) = \ell(w') + 1$.

Exercise 13.7 Show that this pipe dream is indeed reduced. □

We now construct our bijection f. Let $P \in PD(w)$. We will show how to construct $f(P)$ from this pipe dream.

We will illustrate this with pipe dreams for the permutation $w = \overline{216534}$. For it we have $i = 4$ and $j = 6$.

1. The i-th strand intersects with the j-th strand in a box $\boxed{c}$. The i-th strand passes through this crossing from left to right, and the j-th one passes from bottom to top. Replacing this cross by an elbow joint, we get a reduced pipe dream P' for the permutation w'.
2. If the box $\boxed{c}$ is in the i-th row, set $f(P) = P'$. In this case we have $\mathbf{x}^P = x_i \cdot \mathbf{x}^{f(P)}$.

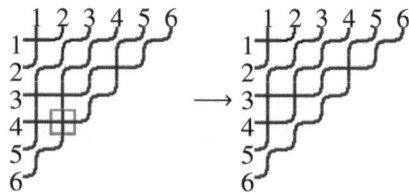

3. Now suppose that the box $\boxed{c}$ is situated in a row $s < i$ (clearly, s cannot be larger than i, since the i-th strand goes only right and up). We removed a cross from row s; now we will add it at some position of the same row.
   The strand i enters $\boxed{c}$ from the left. Take the first elbow occurring on the way from $\boxed{c}$ to the left along i (there may be some crosses between $\boxed{c}$ and this elbow). Such an elbow, which we denote by $\boxed{c'}$, always exists, since $\boxed{c}$ is strictly above the i-th row. The strand i passes through $\boxed{c'}$ from the bottom to the right; denote the other strand, passing through this box from the left to the top, by k.

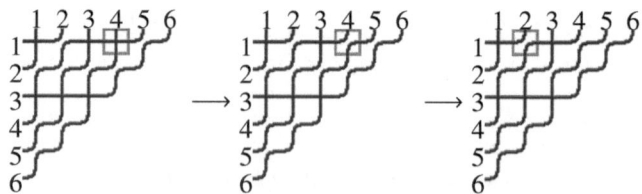

4. If the strands i and k do not intersect, replace the elbow joint in $\boxed{c'}$ by a cross. We obtain a reduced pipe dream P'' of shape $w' \circ (k \leftrightarrow i)$. Since in the box $\boxed{c'}$ strand k passes above strand i, we have $k < i$. Moreover, $\mathbf{x}^{P''} = \mathbf{x}^P$. In this case set $f(P) = P''$.

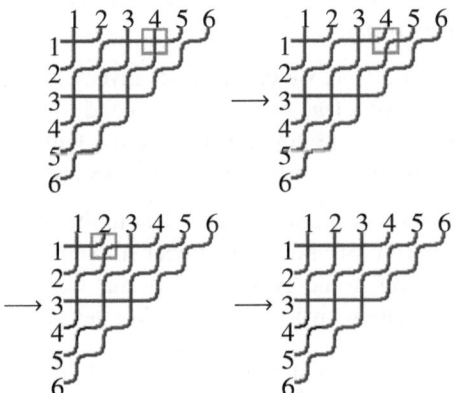

5. Now suppose that strands i and k intersect in the box $\boxed{\mathbf{c''}}$. In the box $\boxed{\mathbf{c'}}$ strand i passes below strand k, thus, if $\boxed{\mathbf{c''}}$ is situated below $\boxed{\mathbf{c'}}$, the strand i enters $\boxed{\mathbf{c''}}$ from the left and leaves from the right. If, on the contrary, $\boxed{\mathbf{c''}}$ is above $\boxed{\mathbf{c'}}$, then the strand i passes through $\boxed{\mathbf{c''}}$ from bottom to top.

Move the cross from $\boxed{\mathbf{c''}}$ into $\boxed{\mathbf{c'}}$. After that the strand i passes through $\boxed{\mathbf{c''}}$ from left to top. Now let us change the notation: let $\boxed{\mathbf{c}} = \boxed{\mathbf{c''}}$, denote the current pipe dream by P', return to Step 2 and proceed.

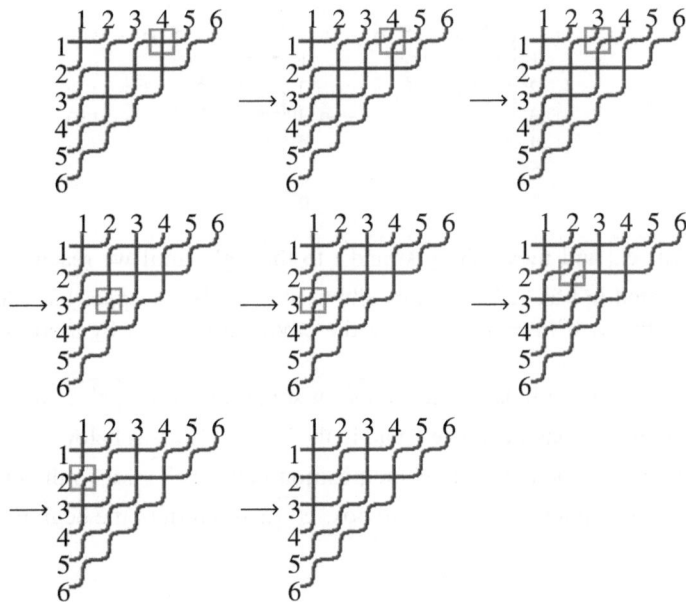

This sequence of operations is finite. Indeed, consider Step 5. If $k < i$, the box $\boxed{\mathbf{c''}}$ is above the box $\boxed{\mathbf{c'}}$, and we move the crossing of these two strands down. If $k > i$, then $\boxed{\mathbf{c''}}$ is below $\boxed{\mathbf{c'}}$, and we move the crossing of these strands up. But every pair of strands intersects at most once, and we cannot move this crossing up or down infinitely many times.

To show that f is a bijection, we construct its inverse. Let $P' \in \mathrm{PD}(w')$.

Since for $m > j$ we have $w(m) > w(i) > w(j)$ and $w'(m) > w'(j) > w'(i)$, the strands with numbers $m > j$ are not allowed to cross with strand i. Moreover, they cannot "nearly meet" this strand, because they are always separated by j.

1. Consider the first elbow joint on strand i; denote the box containing it by $\boxed{c}$. It is located in the i-th row. Strand i is the upper strand passing through $\boxed{c}$; denote the number of the lower strand by $m > i$ (note that we have $m \le j$).
2. If $m = j$, replace the elbow by a cross in $\boxed{c}$. We get the pipe dream $f^{-1}(P')$.

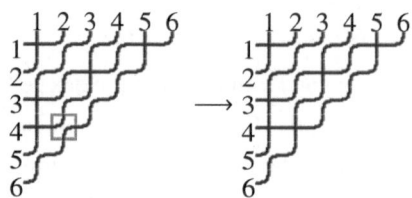

3. If $m < j$, strand m intersects strand i in a box denoted by $\boxed{c'}$ (situated above $\boxed{c}$). Strand i exits from $\boxed{c'}$ on the right. Move the cross from $\boxed{c'}$ into $\boxed{c}$. Strand i still exits $\boxed{c'}$ from the right.

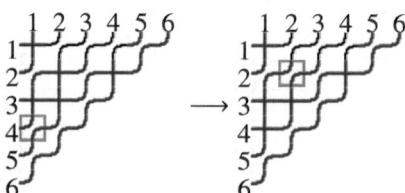

4. Start from $\boxed{c'}$ and move along strand i to the right until we reach the nearest elbow, located in box $\boxed{c''}$. For this elbow, strand i is the upper one; denote the lower one by m. If $m = j$, replacing the elbow in this box by a cross gives us $f^{-1}(P')$.

   If $j > m > i$, the strands m and i cross at some box above $\boxed{c''}$. And if $m < i$, then strand m first passed above i, and in the box $\boxed{c''}$ passes below i. This means that these two strands cross at some position below $\boxed{c''}$. Move the cross from this box (denote it again by $\boxed{c'}$) into box $\boxed{c''}$ and return to the beginning of this step.

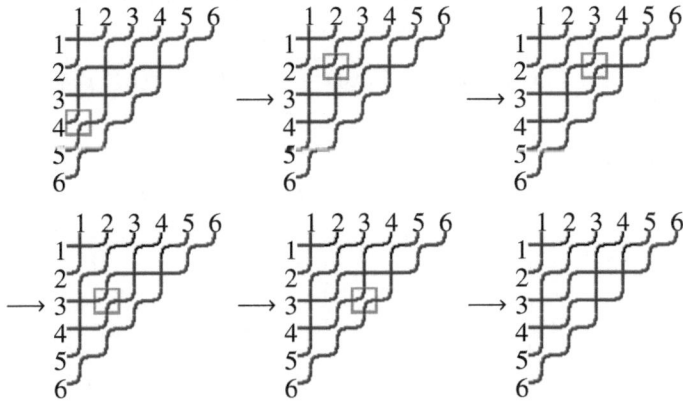

This algorithm moves crosses up for the pairs of strands such that $m < i$; if $m > i$, it moves crosses down. This means that this algorithm stops after a finite number of steps.

Now let $P' \in PD(w' \circ (k \leftrightarrow i))$. Strands i and k cross in a box $\boxed{c'}$. Replace the cross in this box by an elbow; we get a pipe dream of shape w'. Proceeding similarly to the previous algorithm, we can construct $f^{-1}(P')$.

This means that "Schubert polynomials defined in a combinatorial way" $\mathfrak{P}_w$ satisfy the Lascoux transition formula. Now order permutations $w \in S_n$ according to their length, and introduce the reverse lexicographic ordering (of one-line notations) on permutations of the same length. Proceeding by induction on this order, we immediately see that $\mathfrak{P}_w = \mathfrak{S}_w$. $\square$

13.3 Problems

13.1 Let $w \in S_n$ be an arbitrary permutation. What is the relation between the sets of pipe dreams $PD(w)$ and $PD\left(w^{-1}\right)$?

13.2 Using the Bergeron–Billey–Fomin–Kirillov theorem, compute $\mathfrak{S}_{\overline{15423}}$.

13.3 Using the Bergeron–Billey–Fomin–Kirillov theorem, show that

(a) $\mathfrak{S}_{\overline{1,2,\ldots,k-1,k+1,\ldots,n,k}} = e_{n-k}(x_1, x_2, \ldots, x_{n-1})$.
(b) $\mathfrak{S}_{\overline{1,2\ldots,k-1,n,k,k+1,\ldots,n-1}} = h_{n-k}(x_1, x_2, \ldots, x_k)$.

13.4 Let $w_{1,n} = \overline{1, n+1, n, \ldots, 3, 2} \in S_{n+1}$.

(a) Show that the number of reduced pipe dreams of shape $w_{1,n}$ equals the n-th Catalan number C_n.
(b) Construct an explicit bijection between pipe dreams of shape $w_{1,n}$ and triangulations of an $(n+2)$-gon.

(c) Define the q-*Catalan numbers*[2] using the recurrence relation:

$$C_0(q) = 1; \qquad C_n(q) = \sum_{k=0}^{n-1} q^k C_{n-k-1}(q) C_k(q).$$

Show that

$$\mathfrak{S}_{w_{1,n}}(1, q, \dots, q^n) = q^{\binom{n}{3}} C_n(q).$$

---

[2] These are the so-called Carlitz–Riordan q-Catalan numbers, one of several possible q-deformations of the Catalan sequence. It does not coincide with the "obvious" q-deformation $\frac{1}{[n+1]_q} \begin{bmatrix} 2n \\ n \end{bmatrix}$.

Chapter 14
Properties of Pipe Dreams

14.1 Bottom Pipe Dreams

Given a permutation $w \in S_n$, how do we draw a pipe dream (at least one of them) with such a shape? A natural idea would be to arrange crosses and elbow joints in the simplest possible way. For example, we can place all crosses into left-adjusted boxes: i.e., in each row we have a sequence of crosses followed by a sequence of elbow joints.

Definition 14.1 A *bottom pipe dream* is defined by the following property: in each of its rows, any cross is located to the left of any elbow.

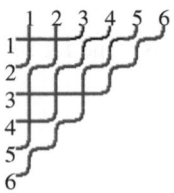

Fig. 14.1: Bottom pipe dream of shape $w = \overline{316425}$

Exercise 14.2 Prove that any bottom pipe dream is reduced.

Consider a bottom pipe dream P with L_i crosses in its i-th row and find its shape w. Each strand i passes horizontally exactly through L_i crossings. This means that there are exactly L_i strands with numbers $j > i$ that cross the i-th strand. Equivalently, we have $L_i = |\{j > i \mid w(j) < w(i)\}|$. So $L = (L_1, L_2, \ldots, L_n)$ is the *Lehmer code* of w.

We obtain a recipe for finding one of the pipe dreams of a given permutation $w \in S_n$: compute the Lehmer code for this permutation and draw the bottom pipe dream with L_i crosses in the i-th row.

E. Smirnov, A. Tutubalina, *Symmetric Functions: A Beginner's Course*,
Moscow Lectures 10, https://doi.org/10.1007/978-3-031-50341-2_14

Exercise 14.3 Draw the bottom pipe dream for the permutation $w = \overline{412653}$ and check that its shape is indeed equal to w.

14.2 The Pipe Dream Graph

As we mentioned in the previous chapter, certain pipe dreams of a given shape $w \in S_n$ can be transformed into each other by moving a cross. If strands i and j intersect in a box $\boxed{c}$ and "nearly meet" in a box $\boxed{e}$, we can move the cross from $\boxed{c}$ to $\boxed{e}$; this operation preserves the shape of pipe dreams.

We can define the *pipe dream graph* for a given permutation w as follows: its vertices are indexed by the set PD(w) (i.e. by pipe dreams of shape w). Two pipe dreams are connected by an edge if and only if they are obtained from each other by moving exactly one cross in the described way.

Exercise 14.4 Draw this graph for $w = \overline{1432}$.

It is natural to ask whether this graph is connected. In other words, is it true that any pipe dreams of the same shape can be brought into each other by consecutive moving of several crosses? It turns out that the answer is positive. Moreover, we can restrict ourselves to a smaller subset of allowed moves.

Definition 14.5 Take a pipe dream with a pair of consecutive elements ⌐⊦, with several (possibly zero) pairs of crosses ⊦⊦ under them, followed by a pair of two elbows. Then the transformation

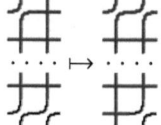
        (dots stand for several rows of crosses)

is called a *ladder move*.

Proposition 14.6 *Every pipe dream can be brought to a bottom pipe dream by a sequence of ladder moves.*

Proof Consider a pipe dream P. If it is not a bottom one, take the lowest row containing an elbow and a cross to the right of it. In this row, take the leftmost pair ⌐⊦.

Under this pair we have several (possibly none) pairs ⊦⊦. Under this column of double crosses we cannot have a pair ⌐⊦, since we already took the lowest of such pairs. Neither can we have a pair ⊦⌐: a reduced pipe dream cannot contain fragments of shape

We are left with a unique possibility: this pair consists of two elbow joints. This means we can apply a ladder move, bringing one cross down.

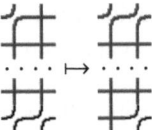

We can continue this procedure until we obtain a bottom pipe dream. Indeed, at each step the sum of column numbers for all crosses decreases (one cross is moved to the column to the left of it). This means that this process will terminate. □

Exercise 14.7 Bring the following pipe dream to a bottom one by a sequence of ladder moves.

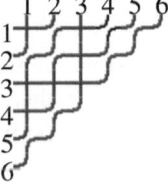

The following statement is immediate.

Corollary 14.8 *The pipe dream graph is connected.*

Proof Indeed, any ladder move corresponds to an edge in the pipe dream graph, and any pipe dream can be brought to the bottom one by a sequence of ladder moves. □

14.3 Schubert Polynomials Form a Basis in the Polynomial Ring

The construction of bottom pipe dreams almost immediately implies that Schubert polynomials form a basis in the polynomial ring. Before we proceed, let us introduce the formalism of symmetric groups of infinitely many variables.

Definition 14.9 Consider the sequence of embeddings of permutation groups:

$$S_1 \hookrightarrow S_2 \hookrightarrow \cdots \hookrightarrow S_n \hookrightarrow S_{n+1} \hookrightarrow \ldots$$

Here $S_n \hookrightarrow S_{n+1}$ as the stabilizer of element $n + 1$. This allows us to define the group $S_\infty = \bigcup_{n=1}^\infty S_n$ of *finitary* permutations of $\mathbb{Z}_{>0}$ (i.e., any permutation $w \in S_\infty$ permutes only finitely many elements).

Exercise 14.10 Prove that any finitary sequence of nonnegative integers (i.e., any sequence with finitely many nonzero elements) $L = (L_1, L_2, \ldots)$ is the Lehmer code of a unique permutation $w \in S_\infty$.

As we saw in Exercise 12.12, Schubert polynomials $\mathfrak{S}_w$ do not depend on the symmetric group containing w. So we may think of Schubert polynomials as elements of the polynomial ring $\mathbb{Z}[x_1, x_2, \ldots]$ in countably many variables; they are indexed by (finitary) permutations $w \in S_\infty$.

Proposition 14.11 *Schubert polynomials $\mathfrak{S}_w$, where $w \in S_\infty$, form a basis in $\mathbb{Z}[x_1, x_2, \dots]$.*

Proof We already know that $\mathfrak{S}_w$ are linearly independent (Exercise 12.1). It remains to prove that any polynomial can be expressed as a linear combination of Schubert polynomials.

The Bergeron–Billey–Fomin–Kirillov theorem (Theorem 13.5) claims that any Schubert polynomial can be obtained as the sum over the set PD(w) of pipe dreams of a given shape:

$$\mathfrak{S}_w = \sum_{P \in \mathrm{PD}(w)} \mathbf{x}^P.$$

Note that the lowest term of a Schubert polynomial (with respect to the lexicographic ordering $x_1 > x_2 > \cdots$) is obtained only from the bottom pipe dream. This implies the following two statements.

- The coefficient in front of the lowest monomial of $\mathfrak{S}_w$ is equal to 1.
- For each monomial $\mathbf{x}^a = x_1^{a_1} x_2^{a_2} \dots x_k^{a_k}$ there exists a Schubert polynomial $\mathfrak{S}_w$ such that its lowest term equals $\mathbf{x}^a$. (The permutation w is defined by its Lehmer code a.)

Now we can prove the required statement by induction on the lowest term. Let $f \in \mathbb{Z}[x_1, x_2, \dots]$ be a homogeneous polynomial of degree n, and let $c_a \mathbf{x}^a$ be its lowest term. Note that the number of monomials of degree n that are lexicographically higher than $\mathbf{x}^a$ is finite.

Find a permutation $w \in S_\infty$, such that the lowest monomial in $\mathfrak{S}_w$ equals $\mathbf{x}^a$. Then either $f = c_a \mathfrak{S}_w$, or the lowest term of the polynomial $g = f - c_a \mathfrak{S}_w$ (denote it by $c_b \mathbf{x}^b$) is lexicographically higher than $\mathbf{x}^a$. By the induction hypothesis, g is a linear combination of Schubert polynomials. Hence f also can be expressed as such a linear combination.

This proves the statement of the theorem for homogeneous polynomials, and any inhomogeneous polynomial is a sum of homogeneous ones. This concludes the proof.   □

Exercise 14.12 Express $f(x_1, x_2, x_3) = x_1^3 + 2x_2^2 x_3$ as a linear combination of Schubert polynomials.

Remark 14.13 Since Schubert polynomials form a basis of the polynomial ring, it would be interesting to compute its structure constants, i.e. the coefficients c_{wv}^u in the expansion

$$\mathfrak{S}_w \mathfrak{S}_v = \sum_u c_{wv}^u \mathfrak{S}_u.$$

Monk's rule and (as we will see in the next section) the Littlewood–Richardson rule are particular cases of this problem.

Unfortunately, there is no combinatorial *manifestly positive* rule for computing the coefficients c_{wv}^u. However, there is a description of Schubert polynomials in terms of algebraic geometry (which is beyond the scope of this book), allowing us to claim that these coefficients are nonnegative. We refer the reader to [Man98] for the details.

14.4 Schubert Polynomials of Grassmannian Permutations

Recall that a permutation is called k-*Grassmannian* if it has a unique descent at the position k. Exercise 12.3 claims that Schubert polynomials of such permutations are symmetric polynomials in the variables $x_1, \ldots, x_k$. Quite miraculously, it turns out that these symmetric polynomials are nothing but Schur polynomials.

Let $w \in S_\infty$ be a k-Grassmannian permutation. Its Lehmer code $L(w)$ is a weakly increasing sequence of length k (cf. Exercise 11.4):

$$L_k(w) \geq L_{k-1}(w) \geq \cdots \geq L_1(w)$$

and

$$L_{k+1}(w) = \cdots = L_n(w) = 0.$$

We obtain a bijection between k-Grassmannian permutations w and partitions $\lambda(w) = (L_k(w), \ldots, L_1(w))$ of length not exceeding k.

Exercise 14.14 Find the partition corresponding to the permutation $w = \overline{135624}$.

Proposition 14.15 *The Schubert polynomial $\mathfrak{S}_w$ for a k-Grassmannian permutation w is equal to the Schur polynomial $s_\lambda(x_1, \ldots, x_k)$.*

Proof Construct a bijection between pipe dreams of shape w and semistandard Young tableaux of shape λ with entries not exceeding k.

Consider the bottom pipe dream of shape w. The crosses in its row form a Young diagram $\lambda(w)$ flipped upside down, with the last (and longest) row of crosses at position k.

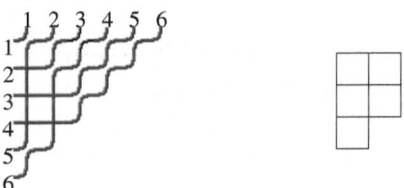

Fig. 14.2: Bottom pipe dream for $w = \overline{135624}$ and Young diagram $\lambda(w)$

Now take an arbitrary pipe dream of shape w and bring it to the bottom one by a sequence of ladder moves.

Exercise 14.16 Show that all these moves have the following form:

(it may be easier to prove the converse statement, lifting crosses of the bottom pipe dream). $\qquad\square$

For each cross, let us keep track of the number of the row it initially came from. If a cross in the bottom pipe dream was in the i-th row, put the number $k - i + 1$ into the corresponding box of the Young tableau.

Fig. 14.3: Pipe dream and the corresponding Young tableau

Exercise 14.17 Show that the result is indeed a semistandard Young tableau.

This map is a bijection: given a Young tableau, we can lift each cross of the bottom pipe dream into the corresponding line. For this, we lift the crosses one by one, proceeding from top to bottom and reading each row from right to left.

Exercise 14.18 Show that the crosses "do not overlap", and each cross can be lifted to the prescribed position.   □

This gives a bijection $f \colon \mathrm{PD}(w) \to \mathrm{SSYT}_\lambda(k)$ between pipe dreams and semistandard Young tableaux. The cross in the i-th row of the pipe dream corresponds to the number $k - i + 1$ in the Young tableau. This means that the corresponding monomial $\mathbf{x}^P$ in the Schubert polynomial and $\mathbf{x}^{f(P)}$ in the Schur polynomial are obtained from one another by a renumbering of variables in the opposite order: $x_i \mapsto x_{k-i+1}$. This means that

$$\mathfrak{S}_w(x_1, \dots, x_k) = s_\lambda(x_k, \dots, x_1).$$

But Schur polynomials are symmetric, and hence

$$\mathfrak{S}_w(x_1, \dots, x_k) = s_\lambda(x_1, \dots, x_k).$$ □

14.5 Problems

14.1 Draw the pipe dream graph for the permutation $w = \overline{15342}$.

14.2 Let $f \in \mathbb{Z}[x_1, x_2, \dots]$ be a polynomial that is symmetric in the variables x_i, x_{i+1}. Show that its Schubert expansion contains only $\mathfrak{S}_w$ such that $w(i) < w(i+1)$.

14.3 (a) What is the relation between the bottom pipe dreams of the permutations w and w^{-1}?

    (b) Let w be a *bigrassmannian* permutation, i.e. w and w^{-1} are both Grassmannian. What can be said about its bottom pipe dream?

14.4 Let $w \in S_n$ be a k-Grassmannian permutation, and let λ be the corresponding partition (cf. Exercise 11.4). Let T be a semistandard Young tableau of shape λ, filled by numbers from 1 to k. Moreover, denote the pipe dream corresponding to T by P_T (cf. Proposition 14.15), and let a_T be the D-dense array corresponding to T.

    (a) Show that the following are equivalent:
- each occurrence of $i + 1$ in T is located strictly to the right of each occurrence of i;
- each cross in the $(k - i)$-th row of P_T is located strictly to the right of each cross in the $(k - i + 1)$-th row;
- the stable matching between the i-th and $(i + 1)$-th columns of the array a_T is empty (i.e., all balls are free).

    (b) If each $i + 1$ in T is located strictly to the right of each i, we can replace the leftmost occurrence of $i + 1$ by i and still obtain a semistandard Young tableau. Describe the corresponding operations for the pipe dream P_T and the array a_T.

14.5 *Double Schubert polynomials* are defined as follows. Assign to a cross in the i-th row and j-th column the weight $x_i + y_j$. Then we have

$$\mathfrak{S}_w(\mathbf{x}, \mathbf{y}) = \sum_{P \in PD(w)} \prod_{\substack{+ \text{ at position} \\ (i,j) \text{ in } P}} (x_i + y_j).$$

    (a) Show that $\mathfrak{S}_w(\mathbf{x}, \mathbf{y}) = \mathfrak{S}_{w^{-1}}(\mathbf{y}, \mathbf{x})$.

    (b) Prove the *Cauchy decomposition formula*:

$$\mathfrak{S}_w(\mathbf{x}, \mathbf{y}) = \sum_{\substack{w=vu \\ \ell(u)+\ell(v)=\ell(w)}} \mathfrak{S}_u(\mathbf{x})\mathfrak{S}_{v^{-1}}(\mathbf{y}).$$

Chapter 15
Problem Set 3

This series of exercises presents another proof of the Bereron–Billey–Fomin–Kirillov theorem on equivalence of algebraic and combinatorial definitions of Schubert polynomials (cf. [BJS93]).

We introduce the following definitions and notation:

Definition 15.1

- The *raising operator* $\uparrow$ acts on the polynomial ring $\mathbb{Z}[\mathbf{x}]$ by a right shift of all variables:

$$\uparrow(x_i) = x_{i+1}.$$

- Let $w \in S_n, w(1) = 1$. The *lowering operator* acts on w as follows:

$$\downarrow(w) = \overline{w(2) - 1, w(3) - 1, \ldots, w(n) - 1} \in S_{n-1}.$$

- Let $Q = s_{i_1} s_{i_2} \ldots s_{i_k}$ be a reduced word for the permutation $w \in S_n$. Define the action of the raising and the lowering operators on it as follows:

$$\uparrow Q = s_{i_1+1} s_{i_2+1} \ldots s_{i_k+1}.$$

If the letter s_1 does not occur in Q, then

$$\downarrow Q = s_{i_1-1} s_{i_2-1} \ldots s_{i_k-1}.$$

- Denote by $\uparrow^j$ the j-th power of the raising operator $\uparrow$.

Definition 15.2 Let $v \in S_n, w \in S_m$. Then

$$v * w = \overline{v(1) + m, \ldots, v(n) + m, w(1), \ldots, w(m)}.$$

Definition 15.3 Let $w \in S_n$, and let $Q_v = s_{i_1} s_{i_2} \ldots s_{i_k}$ be a reduced word for a permutation $v \in S_n$. The word Q_v is called an *initial word* for w if there exists a reduced word for w starting with Q_v, i.e. $\ell(v^{-1}w) = \ell(w) - \ell(v)$.

If $i_1 > i_2 > \cdots > i_k$, then Q_v is said to be *decreasing*.

E. Smirnov, A. Tutubalina, *Symmetric Functions: A Beginner's Course*,
Moscow Lectures 10, https://doi.org/10.1007/978-3-031-50341-2_15

Let $\mathfrak{S}_w$ be Schubert polynomials defined as in Definition 12.7, and let $\mathfrak{P}_w$ be "combinatorial Schubert polynomials" defined by means of pipe dreams, i.e., the right-hand sides of (13.1).

15.1 (Block decomposition of Schubert polynomials) Let $v_1 \in S_n, v_2 \in S_m$.

   (a) Show that
$$\mathfrak{S}_{v_1 * v_2} = (x_1 \ldots x_n)^m \, \mathfrak{S}_{v_1} \!\uparrow^n \left(\mathfrak{S}_{v_2} \right).$$

   (b) Prove a similar formula for "combinatorial" Schubert polynomials:
$$\mathfrak{P}_{v_1 * v_2} = (x_1 \ldots x_n)^m \, \mathfrak{P}_{v_1} \!\uparrow^n \left(\mathfrak{P}_{v_2} \right).$$

15.2 (Decomposition of Schubert polynomials with respect to the first row of a pipe dream) Show that $\mathfrak{P}_w$ satisfies the following recurrence relation:
$$\mathfrak{P}_w = \sum_{\substack{Q_v \text{ decreasing initial word for } w \\ v(1) = w(1)}} x_1^{|Q_v|} \cdot \uparrow \left(\mathfrak{P}_{\downarrow(v^{-1}w)} \right).$$

Further prove that $\mathfrak{S}_w = \mathfrak{P}_w$ for each $w \in S_n$, by induction on n. The base (when $n = 1$) is obvious. In further exercises we shall assume that the induction hypothesis holds for S_n, and prove it for S_{n+1}.

15.3 Let $i > 1$ and $w \in S_{n+1}$.

   (a) Using the induction hypothesis, show that
$$\partial_i \mathfrak{P}_w = 0, \text{ if } \ell(w s_i) > \ell(w).$$

   (b) Using the induction hypothesis, show that
$$\partial_i \mathfrak{P}_w = \mathfrak{P}_{w s_i}, \text{ if } \ell(w s_i) < \ell(w).$$

15.4 (a) Let $u \in S_n$ and $w = u * 1 \in S_{n+1}$. Show that $\mathfrak{S}_w = \mathfrak{P}_w$.
   (b) Let $w \in S_{n+1}$ and $w(1) \neq 1$. Apply to a suitable $\mathfrak{S}_{u*1}$ a sequence of divided differences $\partial_{i_1} \ldots \partial_{i_k}$, with all $i_j > 1$, and show that $\mathfrak{S}_w = \mathfrak{P}_w$.

15.5 Let $w \in S_{n+1}$ and $w(1) = 1$. Proceeding by induction on $\ell(w)$, show that $\mathfrak{S}_w = \mathfrak{P}_w$. For this, suppose this holds for all permutations of length less than $\ell(w)$.

   (a) Show that $\mathfrak{S}_w - \mathfrak{P}_w$ is independent of $x_2, \ldots, x_n$.
   (b) Compare the coefficients in front of $x_1^{\ell(w)}$ and $x_2^{\ell(w)}$ in $\mathfrak{S}_w$ and $\mathfrak{P}_w$. Deduce that $\mathfrak{S}_w = \mathfrak{P}_w$ for $w \in S_{n+1}, w(1) = 1$, thus completing the proof.

Chapter 16
Hints, Answers, and Solutions

Chapter 1

1.1 Write the system of equations (1.1) in the matrix form and solve it using Cramer's rule.

1.2

(a) Take the logarithmic derivative of $E(t) = \sum e_k t^k = \prod (1 + x_i t)$, using the fact that the logarithm of a product equals the sum of logarithms.

$$\frac{d}{dt} \ln E(t) = \sum_{i=1}^{n} \frac{d}{dt} \ln(1 + x_i t) = \sum_{i=1}^{n} \frac{x_i}{1 + x_i t}$$

$$= \sum_{i=1}^{n} \sum_{k \geq 0} x_i^{k+1} (-t)^k = \sum_{k \geq 1} p_k(\mathbf{x}) (-t)^{k-1}.$$

(b) On the other hand, $\frac{d}{dt} \ln E(t) = \frac{E'(t)}{E(t)}$. Thus,

$$\sum_{k=0}^{\infty} k e_k t^{k-1} = \left(\sum_{k=0}^{\infty} e_k t^k \right) \left(\sum_{k=1}^{\infty} (-1)^{k-1} p_k t^{k-1} \right).$$

Comparing the coefficients in front of t^{k-1} in both sides, we get the desired relations.

(c) Write the relations from (b) as a system of linear equations and solve it using Cramer's rule.

1.3 This polynomial is equal to $E(\omega) E(\overline{\omega})$, where $\omega = -\frac{1}{2} + \frac{\sqrt{3}}{2} i$ is a primitive cubic root of unity.

© The Author(s), under exclusive license to Springer Nature Switzerland AG 2024
E. Smirnov, A. Tutubalina, *Symmetric Functions: A Beginner's Course*,
Moscow Lectures 10, https://doi.org/10.1007/978-3-031-50341-2_16

1.4

(a) Consider a table with k rows and n columns $\begin{pmatrix} x_1 \, x_2 \, \dots \, x_n \\ x_1 \, x_2 \, \dots \, x_n \\ \dots\dots\dots \\ x_1 \, x_2 \, \dots \, x_n \end{pmatrix}$ To get a monomial

appearing in e_λ, we need to pick λ_1 variables from the first row, λ_2 variables from the second row, etc. Their product is equal to $x_1^{i_1} \dots x_n^{i_n}$, where i_1 is the number of variables chosen from the first column, i_2 is the number of variables from the second column, and so on. Replacing the chosen variables by 1 and the remaining variables by 0, we get a 0-1 matrix with the prescribed column and row weights.

This shows that the coefficient in front of $x_1^{i_1} \dots x_n^{i_n}$ in e_λ is equal to the number of such matrices. The desired equality follows from the fact that e_λ is symmetric.

(b) Consider a 0-1 matrix with row weight λ and column weight μ. Move all the 1s to the left; we get a matrix where the 1s form a Young diagram of shape λ. Its column weight equals λ'. Since moving 1s to the left increases the column weight of a matrix, we have $\mu \leq \lambda'$.

(c) If $\mu = \lambda'$, then all 1s were left-adjusted in the initial matrix. Such a matrix is unique.

1.5

(a) The proof is similar to the previous exercise, but now each variable can be taken several times.

(b) $N_{\lambda, (1^m)} = \binom{m}{\lambda_1, \lambda_2, \dots, \lambda_k}$

Chapter 2

2.1

(a) $\prod_{i<j}(x_i - x_j)$;

(b) $a_1 \dots a_n \prod_{i<j}(x_i - x_j)$.

2.2

(a) Multiply the determinant by $\prod_{i,j=1}^{n}(x_i + y_j)$. Subtracting columns from each other, take out the multiple $(y_i - y_j)$ and use skew symmetry.

(b) Use the change of variables $x_k \mapsto x_k^{-1}$.

(c) Subtract rows from each other to get a factor $(b_i - a_j)$.

2.3 Use the previous problems, substituting numerical values instead of variables.

2.5 Expand the determinant $a_{(r+n-1, n-2, n-3, \dots, 2, 1, 0)}$ with respect to the first row.

2.6

(b) Replace each summand $h_{a+k}e_{b-k+1}$ in the right-hand side by $s_{(a+k, 1^{b-k+1})} + s_{(a+k+1, 1^{b-k})}$. We get a "telescopic sum" where all summands except the first one cancel out.

Another method is as follows: use the Jacobi–Trudi formula to write $s_{(a+1,1^b)}$ as a determinant in h_k and expand it with respect to the first row.

2.7 Let $\lambda = (\lambda_1, \ldots, \lambda_m)$ be a partition without zeros at the end. Let us show how to express a Schur polynomial s_λ in e_k using the Pieri formulas. We shall use induction in λ (with respect to the lexicographic ordering). The base is obvious: $s_{(1^k)} = e_k$. Suppose all s_μ with $\mu < \lambda$ are expanded in e_k.

Consider a partition $\nu = \lambda - (1^m) = (\lambda_1 - 1, \lambda_2 - 1, \ldots, \lambda_m - 1)$. The polynomial s_ν can be expanded in e_k. This means that

$$s_\nu \cdot e_m = \sum_{\mu \in \nu \otimes 1^m} s_\mu = s_\lambda + \sum_\mu s_\mu,$$

and all the μ's occurring in the last sum are strictly less than λ (since the maximal diagram in $\mu \in \nu \otimes 1^m$ is obtained by adding a box into each of the first m rows, and this is exactly λ). All s_μ can be expanded in e_k by the induction hypothesis, so s_λ also can be expanded in e_k.

It remains to observe that taking conjugates replaces λ by λ', and e_k by h_k. Hence $s_{\lambda'}$ is expanded in h_k in the same way as s_λ is expanded in e_k. This means that ω indeed brings s_λ to $s_{\lambda'}$.

Chapter 3

3.1 (a) $K_{\lambda\mu} = k + 1$;   (b) $K_{\lambda\mu} = \binom{a+b-c}{b-c}$;   (c) $K_{\lambda\mu} = C_k$ (k-th Catalan number).

3.2 First observe that $K_{\lambda\lambda} = 1$ (there is a unique tableau of shape and weight λ, with the first row filled by 1, the second row filled by 2, etc). Next, if there exists a Young diagram of shape λ and weight μ, all the 1s must be in the first row, all 1s and 2s in the first two rows, etc. This means that $\mu_1 \leq \lambda_1, \mu_1 + \mu_2 \leq \lambda_1 + \lambda_2$, and so on. So if $K_{\lambda\mu} > 0$, we have $\mu \leq \lambda$.

Using induction by λ (with respect to the partial ordering $\leq$), we can express m_λ as linear combinations of Schur polynomials. This proves that Schur polynomials form a basis in Λ_n.

3.3 Construct a Young tableau of shape $\lambda = (\lambda_1, \lambda_2)$ and weight $\mu \leq \lambda$. For this write 1 in $\mu_1 \leq \lambda_1$ boxes of the first row, followed by 2 in μ_2 boxes, then 3 in μ_3 boxes, etc. After exhausting the boxes in the first row, proceed with the second row etc. Since $\mu_i \leq \mu_1 \leq \lambda_1$, we will never obtain two boxes with the same label i standing under one another.

3.4 Using the Pieri formula, express $h_\lambda = h_{\lambda_k} \ldots h_{\lambda_1}$ as a linear combination of Schur polynomials. Start with $h_{\lambda_1} = s_{(\lambda_1)}$, which is a Schur polynomial in its own right. Draw a row of length λ_1 and label its boxes with 1s.

To expand $h_{\lambda_2} s_{(\lambda_1)}$ in Schur polynomials we need to add λ_2 boxes into distinct columns in all possible ways (to get a Young diagram). Label these boxes with 2s.

Then add to these diagrams λ_3 boxes into distinct columns, label them with 3s, and so on.

We finally get a set of Young tableaux of different shapes μ and of fixed weight λ. Each such tableau contributes s_μ into the expansion of h_λ with respect to Schur polynomials, which implies the desired statement.

The second formula is proven similarly, but the usual semistandard Young tableaux should be replaced by transposed (row-strict and column-semistrict) ones.

3.5

(a) Put the weights on horizontal edges along antidiagonals, as shown in the figure below.

x_4	x_5	x_6	x_7
x_3	x_4	x_5	x_6
x_2	x_3	x_4	x_5
x_1	x_2	x_3	x_4

(b) Denote the number of columns in the Young diagram by ℓ. Each column corresponds to its lattice path. Take $\{A_1, \ldots, A_\ell\}$, with $A_i = (-i, i)$, as the initial points of these paths, and $\{B_1, \ldots, B_\ell\}$, with $B_i = (\lambda_i' - i, n + i - \lambda_i')$, as the final ones. Now apply the LGV lemma.

3.7

(a) Take $\{M_0, \ldots, M_{n-1}\}$ as the initial points and $\{M_1, \ldots, M_n\}$ as the final ones. Then a noncrossing set of paths consists of $n - 1$ "one-point" paths $M_i \to M_i$ and one path $M_0 \to M_n$ that does not pass through the remaining points M_i. Apply the LGV lemma, paying attention to the sign of a permutation.

(b) The n-th Catalan number C_n equals one half of the number of paths in an $(n + 1) \times (n + 1)$ square not passing through the points of the diagonal (half of such paths are above the diagonal and the other half are below it).

Chapter 4

4.1

(b) Apply the involution ω acting on the **x**-variables to the relation $\sum_\lambda h_\lambda(\mathbf{x}) m_\lambda(\mathbf{y}) = \sum_\lambda s_\lambda(\mathbf{x}) s_\lambda(\mathbf{y})$.

4.2

(a) Applying the relation

$$(1 - s)^{-1} = \exp\left(s + \frac{s^2}{2} + \frac{s^3}{3} + \cdots\right)$$

to $H(t) = \prod_i (1 - x_i t)^{-1}$, we get

$$H(t) = \exp\left(\sum_{k=1}^{\infty} \frac{p_k t^k}{k}\right) = \sum_{n=0}^{\infty} \frac{1}{n!}\left(\sum_{k=1}^{\infty} \frac{p_k t^k}{k}\right)^n$$

$$= \sum_{n=0}^{\infty} \sum_{k_1,\dots,k_n=1}^{\infty} \frac{p_{k_1} \cdots p_{k_n}}{k_1 \cdot \dots \cdot k_n \cdot n!} t^{k_1 + \dots + k_n}.$$

Now consider the summands with $(k_1, \dots, k_n)$ obtained from a fixed partition λ. The number of nontrivial permutations of its entries $\lambda = (1^{m_1}, 2^{m_2}, 3^{m_3}, \dots)$ equals $\frac{n!}{m_1! \cdot m_2! \cdot m_3! \dots}$.

(b) Use the identity $1 + s = \exp\left(s - \frac{s^2}{2} + \frac{s^3}{3} - \cdots\right)$.

4.4

(a) Apply the involution ω in $\mathbf{x}$-variables to formulas from Exercise 4.3 and derive that

$$\sum_\lambda z_\lambda^{-1}(\omega p_\lambda)(\mathbf{x}) p_\lambda(\mathbf{y}) = \sum_\lambda \varepsilon_\lambda z_\lambda^{-1} p_\lambda(\mathbf{x}) p_\lambda(\mathbf{y}).$$

Now use the linear independence of $p_\lambda(\mathbf{y})$.

(b) Follows from Lemma 4.10.

4.5 First prove this formula for power sums p_λ.

4.6 Start with the case $f = s_\lambda$.

Chapter 5

5.2 The left-hand side is the generating function $E(t)$ evaluated at $x_i = q^i$.

5.3 For each layer of the plane partition take the path bounding the corresponding Young diagram. We get c paths in an $a \times b$ rectangle that can have common segments but never cross each other. Shifting them diagonally, we get a tuple of c nonintersecting paths.

More formally: take $A_i(-i, 1)$ as the initial points and $B_i(a - i, b + i)$ as the final ones. Then the number of paths from A_i to B_j is equal to $\binom{a+b+i-1}{b+j-1}$. According to the LGV lemma, the number of tuples of nonintersecting paths equals the determinant in the right-hand side. It remains to prove that the i-th path in each such family starts with at least $i - 1$ vertical steps.

5.4 Write each binomial coefficient as the ratio of falling factorials, and use the relation $x^{\underline{m+n}} = x^{\underline{m}} \cdot (x - m)^{\underline{n}}$.

5.5 $\prod_{i<j}(j - i) = \prod_{k=1}^{c-1} k^{c-k} = \prod_{j=1}^{c}(j - 1)!$

5.6 Using Exercise 5.5 and the formula from Exercise 5.4, we see that

$$
\mathbb{P}(a, b, c) = \prod_{j=1}^{c} \frac{(a + b + j - 1)^{\underline{b}} \cdot (j - 1)!}{(b + j - 1)!}
$$

$$
= \prod_{j=1}^{c} \frac{(a + b + j - 1)^{\underline{b}}}{(b + j - 1)^{\underline{b}}}
$$

$$
= \prod_{j=1}^{c} \prod_{i=1}^{b} \frac{a + i + j - 1}{i + j - 1}.
$$

Now use the relation

$$
\frac{a + i + j - 1}{i + j - 1} = \frac{a + i + j - 1}{a + i + j - 2} \cdot \frac{a + i + j - 2}{a + i + j - 3} \cdot \ldots \cdot \frac{i + j}{i + j - 1}
$$

to make the formula symmetric in all three variables a, b, and c:

$$
\mathbb{P}(a, b, c) = \prod_{k=1}^{a} \prod_{i=1}^{b} \prod_{j=1}^{c} \frac{i + j + k - 1}{i + j + k - 2}.
$$

5.7 Repeat the computations in Exercises 5.3–5.6 with integers replaced by their q-analogues. Pay attention to the exponent of the common factor q^N.

5.8 Consider the formula

$$
\mathbb{P}_q(a, b, c) = \prod_{i=1}^{b} \prod_{j=1}^{c} \frac{[a + i + j - 1]}{[i + j - 1]}
$$

as b and c tend to infinity. We obtain that

$$
\mathbb{P}_q(a) = \prod_{i=1}^{\infty} \prod_{j=1}^{\infty} \frac{[a + i + j - 1]}{[i + j - 1]}.
$$

Make the change of variables $k = i + j - 1$. In this product each factor $\frac{[a+k]}{[k]}$ occurs exactly k times. So we have

$$
\mathbb{P}_q(a) = \prod_{k=1}^{\infty} \left(\frac{[a + k]}{[k]} \right)^k = \prod_{k=1}^{\infty} \left(\frac{1 - q^{a+k}}{1 - q^k} \right)^k.
$$

As $a \to \infty$, the numerators tend to 1, so the limit is equal to

$$
\mathbb{P}(q) = \prod_{k=1}^{\infty} \frac{1}{\left(1 - q^k\right)^k}.
$$

Chapter 6

6.1 It is clear that for $|\lambda| = n$ with all λ_i odd, the corresponding p_λ belongs to $\Omega^{(n)}$. Let $V^{(n)} \subseteq \Omega^{(n)}$ be their linear span. Show that $V^{(n)} = \Omega^{(n)}$.

Let $f \in \Omega^{(n)}$. Expand f in the basis of power sums:

$$f = \sum_\lambda c_\lambda p_\lambda.$$

Here all $c_\lambda \neq 0$, and the sum is finite. Let f' be the projection of f to $V^{(n)}$, and $\tilde{f} = f - f'$. Then

$$\tilde{f} = \sum_{\substack{\lambda \\ \lambda \text{ has even parts}}} c_\lambda p_\lambda.$$

Show that if $\tilde{f}$ is supersymmetric, then it must be zero. Suppose the contrary and take the smallest even part $2n$ in each λ:

$$\tilde{f} = \sum_{\substack{n,\mu \\ \text{all even } \mu_i \geq 2n}} c_{n,\mu} p_{2n} p_\mu.$$

Let N be the smallest of all n, and let v be the lexicographically lowest partition with $n = N$. Then

$$\tilde{f} = c_{N,v} p_{2N} p_v + \sum_{\substack{n>N \text{ or} \\ n=N; \mu > v \\ \text{all even } \mu_i \geq 2n}} c_{n,\mu} p_{2n} p_\mu.$$

Now compute $\tilde{f}(t, -t, x_3, x_4, \dots)$ and consider the monomial $t^{2N} x_3^{v_1} x_4^{v_2} x_5^{v_3} \dots$. It is easy to see that it can occur only in the first summand. But then the coefficient in front of it is nonzero, and $\tilde{f}(t, -t, x_3, \dots)$ depends on t. Thus $\tilde{f}$ is not supersymmetric, which contradicts our assumption.

So for each $f \in \Omega^{(n)}$ we have $\tilde{f} = 0$, and $V^{(n)} = \Omega^{(n)}$, as desired.

6.2 We shall use the LGV lemma to prove this formula. Consider a dotted vertical line passing through midpoints of segments. Label the horizontal segments in the left-hand part by x_i for the segments on the i-th horizontal line, counted from the bottom (just like in the proof of Theorem 3.4)). In the right-hand part, put x_i on the horizontal segments of the i-th antidiagonal, like in the proof of the second Jacobi–Trudi identity (cf. Exercise 3.5)).

Our paths will go either down or right in the left-hand part of the plane and either up or right in the right-hand part.

For the initial point of a path, take a point A in the left-hand part on the n-th horizontal line such that the distance from the dotted line equals $a + 0.5$. For the terminal point, take a point B in the right-hand part with the distance $b + 0.5$ from the dotted line. Then the sum of monomials over all paths between such points equals $s_{(a|b)}(x_1, \dots, x_n)$. Indeed, the indices along the path first weakly decrease and then strongly increase. This is exactly the behavior of indices in the boxes of a hook Young tableau, read from right to left and then from top to bottom.

A Young tableau decomposed into k hooks corresponds to a collection of k nonintersecting lattice paths. The Giambelli formula now immediately follows from the LGV lemma.

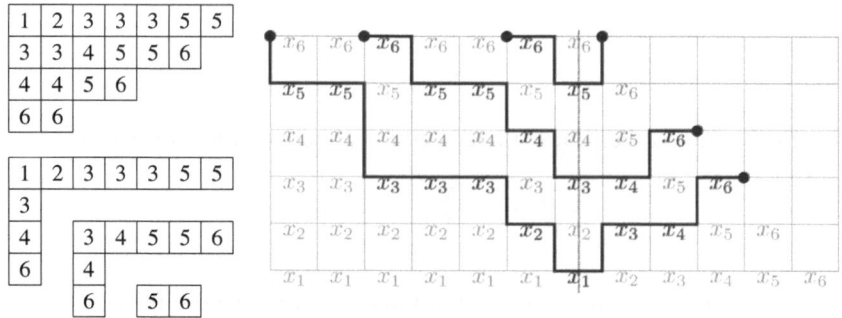

Fig. 16.1: A Young tableau and the collection of lattice paths corresponding to its hook decomposition

6.3 These formulas can be proven by directly applying the LGV lemma.

(a) Assign weights to the edges as in the proof of Littlewood's theorem. Take $A_i(\mu_i - i, 1)$ and $B_i(\lambda_i - i, n)$ as initial and terminal points, respectively.

(b) Assign weights to the edges as in the proof of the second Jacobi–Trudi identity. The initial and terminal points are chosen as $A_i(\mu_i' - i, i - \mu_i')$ and $B_i(\lambda_i' - i, n + i - \lambda_i')$.

6.4

(a) Expand $\tilde{s}_{\lambda/\mu}(\mathbf{x})$ in a basis of Schur polynomials:

$$\sum_{\lambda} \tilde{s}_{\lambda/\mu}(\mathbf{x})s_{\lambda}(\mathbf{y}) = \sum_{\lambda,\nu} \langle \tilde{s}_{\lambda/\mu}(\mathbf{x}), s_{\nu}(\mathbf{x})\rangle s_{\nu}(\mathbf{x})s_{\lambda}(\mathbf{y}).$$

Change the variables $\mathbf{x}$ by $\mathbf{y}$ in the scalar product and use the definition of $\tilde{s}_{\lambda/\mu}(\mathbf{x})$:

$$\sum_{\lambda,\nu} \langle \tilde{s}_{\lambda/\mu}(\mathbf{x}), s_{\nu}(\mathbf{x})\rangle s_{\nu}(\mathbf{x})s_{\lambda}(\mathbf{y}) = \sum_{\lambda,\nu} \langle s_{\mu}(\mathbf{y})s_{\nu}(\mathbf{y}), s_{\lambda}(\mathbf{y})\rangle s_{\lambda}(\mathbf{y})s_{\nu}(\mathbf{x})$$

$$= \sum_{\nu} s_{\mu}(\mathbf{y})s_{\nu}(\mathbf{y})s_{\nu}(\mathbf{x}).$$

Write the common factor $s_{\mu}(\mathbf{y})$ in front of the sum and use Schur's formula:

$$\sum_{\lambda} \tilde{s}_{\lambda/\mu}(\mathbf{x})s_{\lambda}(\mathbf{y}) = s_{\mu}(\mathbf{y})\sum_{\nu} s_{\nu}(\mathbf{y})s_{\nu}(\mathbf{x}) = s_{\mu}(\mathbf{y})\sum_{\nu} h_{\nu}(\mathbf{x})m_{\nu}(\mathbf{y}).$$

(b) Multiply both sides of the equality by $a_\delta(\mathbf{y})$ and expand $\sum_\nu m_\nu(\mathbf{y})$ as the sum of monomials $\mathbf{y}^\alpha$ over all possible finite sequences of nonnegative integers α:

$$\sum_\lambda \tilde{s}_{\lambda/\mu}(\mathbf{x}) a_{\lambda+\delta}(\mathbf{y}) = a_{\mu+\delta}(\mathbf{y}) \sum_\nu h_\nu(\mathbf{x}) m_\nu(\mathbf{y})$$

$$= a_{\mu+\delta}(\mathbf{y}) \sum_{\alpha \in \mathbb{Z}_{\geq 0}^n} h_\alpha(\mathbf{x}) \mathbf{y}^\alpha$$

$$= \left(\sum_{w \in S_n} (-1)^w \mathbf{y}^{w(\mu+\delta)} \right) \left(\sum_{\alpha \in \mathbb{Z}_{\geq 0}^n} h_\alpha(\mathbf{x}) \mathbf{y}^\alpha \right)$$

$$= \sum_{\alpha \in \mathbb{Z}_{\geq 0}^n} h_\alpha(\mathbf{x}) \sum_{w \in S_n} (-1)^w \mathbf{y}^{\alpha+w(\mu+\delta)}.$$

(c) Comparing the coefficients in front of $\mathbf{y}^{\lambda+\delta}$ in both sides, we obtain that

$$\tilde{s}_{\lambda/\mu}(\mathbf{x}) = \sum_{w \in S_n} (-1)^w h_{\lambda+\delta-w(\mu)-w(\delta)} = \det \left(h_{\lambda_i - \mu_j - i + j} \right)_{i,j=1}^n,$$

as desired.

Chapter 7

7.1

(a)
0	0	1
0	2	2
3	1	4

(b)
0	0	1
0	2	1
3	1	1

(c)
0	0	1
0	3	2
6	3	3

(d)
0	0	1
0	2	3
3	4	5

7.2 $(2n - 1, 2n - 3, \ldots, 3, 1)$.

7.3 This inequality means exactly that each ball in row $(j + 1)$ has a matching ball in row j.

7.4 It is enough to consider two rows (without loss of generality suppose that these are rows 1 and 2) and show that the central symmetry brings the matching between these rows into the symmetric matching.

Proceed by induction on the number of pairs in the matching. Take the leftmost matched ball in row 2; denote it by $\mathbf{b}$. Suppose it is located in box $(a + k, 2)$ and matched to a ball $\mathbf{b}'$ in box $(a, 1)$. Note there are no balls in the first row between $\mathbf{b}$ and $\mathbf{b}'$.

After applying central symmetry the ball **b** is located in box $(b, 1)$, while **b'** is located in $(b + k, 2)$, with no balls in the second row between them. We proceed with the balls in the second row from left to right, and at the moment we reach **b'**, the ball **b** will be still free. So these two balls are being matched to each other.

Remove the pair **b**-**b'**. All the other pairs in the matching remain intact; by the induction hypothesis, they are preserved by the symmetry.

7.5

(a) Since condensation operations leave the array shape unchanged, we can assume that a is bidense of shape $\lambda = (\lambda_1, \ldots, \lambda_k)$. Apply symmetry to this array and start condensating it.

It is easy to bring a^S to a diagonal array with $\lambda_k \leq \cdots \leq \lambda_2 \leq \lambda_1$ balls on the diagonal. Applying $L_{k-i}^{\lambda_i - \lambda_{i+1}} D_{k-i}^{\lambda_i - \lambda_{i+1}}$ to it moves $\lambda_i - \lambda_{i+1}$ balls from box $(k - i + 1, k - i + 1)$ to box $(k - i, k - i)$. So this interchanges λ_i and λ_{i+1} in the row (and column) weight of this array.

Applying such operations, we can arrange all λ_i in weakly decreasing order, bringing a^S again to the bidense array a.

(b) Let a be a D-dense array. Exercise 7.4 implies that $Da^S = (Ua)^S$. Thus $\sigma\left(Da^S\right) = D(U(a)) = a$, so σ is an involution.

Chapter 8

8.1

(a)

| 1 | 2 | 3 | | | n |

and

| 1 | 2 | 3 | | | n |

;

(b)

1	i
2	
n	

and

1	i
2	
n	

;

(c)

1	3	6
2	5	9
4	7	
8		

and

1	3	4
2	7	9
5	8	
6		

.

8.2

(a) $w = \overline{198765243}$;

(b) $w = \overline{794628315}$.

8.3 Apply the fiber product theorem to:

(a) symmetric $n \times n$ arrays with exactly one ball in each row and each column;

(b) $m \times n$ arrays with exactly one ball in each row;

(c) $m \times k$ arrays with exactly n balls.

8.4

(a) Apply to a pairs of operations $L_i D_i$ preserving its symmetry. Show that each such pair either does not change the parity of $a(k, k)$, or decreases $a(i+1, i+1)$ by one and decreases $a(i, i)$ by one (but this is the case only if $a(i+1, i+1) = a(i, i) + 1$).

(b) Apply an argument similar to the proof of Schur's formula to symmetric arrays with an even number of balls in each box of their diagonal.

(c) Use the fact that $\sum_k e_k = \prod_i (1 + x_i)$ and apply the involution ω to this identity. The first identity also follows from the Pieri formula (each partition is obtained from an even one by adding at most one box to each row).

Chapter 9

9.1

(a) $s_{(2,1)} \cdot s_{(1,1)} = s_{(3,2)} + s_{(3,1,1)} + s_{(2,2,1)} + s_{(2,1,1,1)}$;

(b) $s_{(2,1)} \cdot s_{(2,1)} = s_{(4,2)} + s_{(4,1,1)} + s_{(3,3)} + 2s_{(3,2,1)} + s_{(3,1,1,1)} + s_{(2,2,2)} + s_{(2,2,1,1)}$;

(c) $s_{(2,2)} \cdot s_{(2,1)} = s_{(4,3)} + s_{(4,2,1)} + s_{(3,3,1)} + s_{(3,2,2)} + s_{(3,2,1,1)} + s_{(2,2,2,1)}$;

(d) $s_{(3,1)} \cdot s_{(2,1)} = s_{(5,2)} + s_{(5,1,1)} + s_{(4,3)} + 2s_{(4,2,1)} + s_{(4,1,1,1)} + s_{(3,3,1)} + s_{(3,2,2)} + s_{(3,2,1,1)}$.

9.4

(a) The right-hand side is an alternating sum of Schur polynomials. Each of them is computed as the sum of the monomials corresponding to hook-shaped Young tableaux. Consider the set of such tableaux. It contains tableaux of shape (k) filled by identical entries. They provide summands x_i^k in each power sum.

Construct an involution on the set of remaining tableaux as follows. For a hook, consider its rightmost entry r and its lowest entry d. If $d \geq r$, move the lowest box to the right, and if $r > d$, do the opposite: move the rightmost box to the bottom of the hook.

Obviously, it is an involution without fixed points. The tableaux related by it correspond to the same monomial; since the number of rows is changed by 1, they occur with opposite signs and cancel out. The only surviving monomials are x_i^k.

(b) Using the Littlewood–Richardson rule, multiply s_λ by $s_{(a,1^{b-1})}$. For this, add to λ the following boxes: a labeled by one, and one box for each of the labels from 2 to b. These boxes must form a skew Young tableau and a Yamanouchi text at the same time.

This condition can be restated as follows: each row should look like

or ,

and each column contains either several consecutive integers or a sequence of consecutive integers preceded by 1.

This implies that the added boxes form one or several ribbons along the edge of the initial diagram.

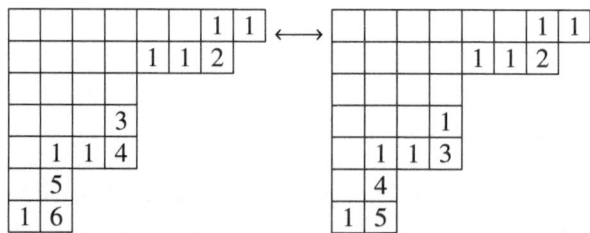

If this is a single ribbon, it corresponds to a polynomial $\pm s_\mu$ from the right-hand side of the Murnaghan–Nakayama rule (make sure that the sign is correct).

The set of skew Young tableaux consisting of several disconnected ribbons admits the following involution. Take the top (rightmost) box of the lowest ribbon. If its label is greater than 1, let us replace it by 1 and decrease by one each label in its ribbon that is greater than 1. If this box is labeled by 1, then denote the maximal number in the remaining rows by $k - 1$. Replace this 1 by k and increase by 1 all the remaining entries in this ribbon.

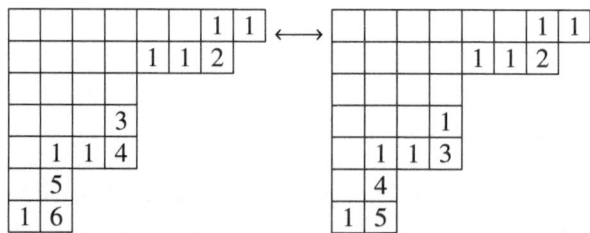

One of these tableaux is obtained from multiplying s_λ by $s_{(a,1^{b-1})}$, while the second one is obtained from multiplying it by $s_{(a-1,1^b)}$. These Schur polynomials appear with opposite signs, so the corresponding polynomials s_μ cancel out.

(c) Let $\ell(\lambda) = n$. Consider all fillings T of the Young diagram of shape μ by $1, \ldots, n$ satisfying the following conditions:

- the labels weakly increase both along rows and columns;
- the number i occurs λ_i times;
- the boxes labeled with i form a ribbon (a connected subset not containing 2×2 squares). Denote by $\mathrm{ht}(i)$ the height of this ribbon *minus one*.

Let $\mathrm{ht}(T) = \mathrm{ht}(1) + \mathrm{ht}(2) + \cdots + \mathrm{ht}(n)$. Then $\chi_\lambda^\mu = \sum(-1)^{\mathrm{ht}(T)}$, where the sum is taken over all such fillings.

(d) Let us compute the coefficient in front of $s_{(5,5,4,2)}$ in $p_{(5,4,4,2,1)}$. List all the fillings that satisfy the above rules:

1)
1	1	1	1	1
2	2	2	2	5
3	3	3	3	
4	4			

2)
1	1	1	1	1
2	2	2	3	5
2	3	3	3	
4	4			

3)
1	1	1	1	1
2	2	2	4	4
2	3	3	5	
3	3			

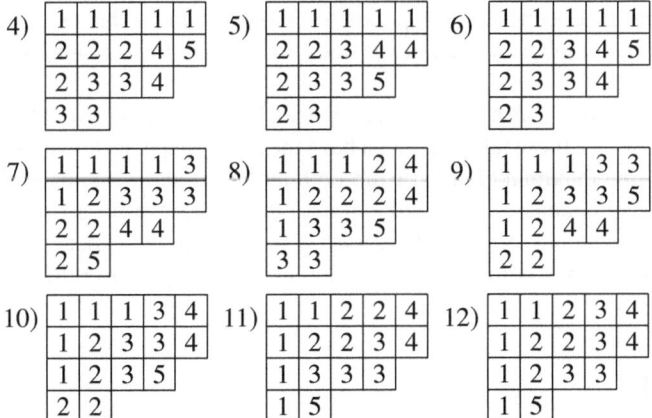

Now compute $\mathrm{ht}(T)$ for each filling:

$$\mathrm{ht}(T_1) = 0+0+0+0+0 = 0 \qquad \mathrm{ht}(T_2) = 0+1+1+0+0 = 2$$
$$\mathrm{ht}(T_3) = 0+1+1+0+0 = 2 \qquad \mathrm{ht}(T_4) = 0+1+1+1+0 = 3$$
$$\mathrm{ht}(T_5) = 0+2+2+0+0 = 4 \qquad \mathrm{ht}(T_6) = 0+2+2+1+0 = 5$$
$$\mathrm{ht}(T_7) = 1+2+1+0+0 = 4 \qquad \mathrm{ht}(T_8) = 2+1+1+1+0 = 5$$
$$\mathrm{ht}(T_9) = 2+2+1+0+0 = 5 \qquad \mathrm{ht}(T_{10}) = 2+2+2+1+0 = 7$$
$$\mathrm{ht}(T_{11}) = 3+1+1+1+0 = 6 \qquad \mathrm{ht}(T_{12}) = 3+2+2+1+0 = 8$$

We get 7 even and 5 odd numbers, hence $\chi^{(5,5,4,2)}_{(5,4,4,2,1)} = 2$.

Chapter 10

10.2 This follows from the fact that the Schensted insertion is invertible.

10.3 Tableau P is obtained as the result of consecutive insertion of terms of the sequence $j_1, \ldots, j_m$. The "counting tableau" Q is filled not by consecutive integers, as in the case of Robinson correspondence, but rather by $i_1, \ldots, i_m$. This correspondence is bijective, since the Schensted insertion is invertible.

10.5 This follows from the Knuth correspondence and the fact that $s_\lambda(x)s_\lambda(y)$ equals the sum of monomials corresponding to pairs of semistandard tableaux of shape λ.

10.6 (a) Using the plactic relations, move the letter x in $m(T)x$ as far to the left as possible. Note that if w_1 is the word corresponding to the first row of the tableau and y is its leftmost element that is greater than x, then $w_1 x \sim y w_1'$, where w_1' is obtained from w_1 by replacing y by x. Then use induction on the number of rows.

(b) A word can be transformed into a tableau word by consecutive insertion of its letters into the empty tableau.

10.7 (a) Show that applying one plactic relation to a sequence does not affect ℓ_i.

(b) Each weakly increasing subsequence of a tableau word corresponds to the collection of boxes with strictly increasing column numbers. Thus picking a set of

k nonintersecting sequences corresponds to taking at most k boxes in each of the columns. But the number of such boxes does not exceed the total length of the first k rows.

(c) Suppose there exists a Knuth equivalence class containing two tableaux T and T'. According to (b), these tableaux must have the same shape. Remove the rightmost occurrence of the maximal element from both tableaux; show that these words will remain Knuth equivalent, and use induction on the number of boxes in T.

(d) Each Young tableau with at least $pq + 1$ boxes contains either a row of length not less than $p + 1$, or a column of length not less than $q + 1$ (or both).

Chapter 11

11.2 Consider a pair of integers $i < j$. It forms an inversion in exactly $n!/2$ permutations from S_n. So the total number of inversions is $\frac{n! \cdot n(n-1)}{4}$.

11.5

(a) The Rothe diagram of a 132-avoiding permutation consists of a unique component (Young diagram in the upper left corner). Its Lehmer code is a partition (nonincreasing sequence).

(b) The number of 132-avoiding permutations equals the n-th Catalan number C_n.

11.6

(a) The "if" part is obvious. To prove the "only if" part, consider a 321-avoiding permutation such that $i_a < i_b, w(i_a) \geq i_a, w(i_b) \geq i_b$, but $w(i_a) > w(i_b)$. Suppose that $w(i_b) > i_b$. Then the interval from i_b to n contains more integers than the interval from $w(i_b)$ to n. Hence there exists a $t > i_b > i_a$ such that $w(t) < w(i_b) < w(i_a)$, which is a contradiction. If $w(i_b) = i_b$, then $w(i_a) > i_a$, and so on. The part about j's is proven similarly.

(b) The "only if" part: consider two reduced words for w that are obtained from each other by a braid relation. Then in one of them we have a triple of consecutive letters $s_i s_{i+1} s_i$. Consider the wiring diagram for this word. In its corresponding part the order of these three strands is reversed. They have no other intersections, so the whole permutation also reverses the order of a certain triple of elements. So it is not 321-avoiding.

The "if" part: suppose that w is not 321-avoiding. Then for some triple $i < j < k$ we have $w(i) > w(j) > w(k)$. Select j and $i < j$ in such a way that $w(i)$ is maximal, and $k > j$ in such a way that $w(k)$ is minimal. Then for each m between i and j the pair (i, m) is an inversion. Multiply w from the right by $s_i, s_{i+1}, \ldots, s_{j-2}$; the length of the permutation decreases at each step. Then multiply w from the right by $s_{k-1}, s_{k-2}, \ldots, s_{j+1}$; each of these multiplications also decreases the length of the permutation.

We finally get a permutation v such that

$$v(j - 1) = w(i), \quad v(j) = w(j), \quad v(j + 1) = w(k).$$

There exists a reduced word for v ending with $\ldots s_{j-1}s_j s_{j-1}$. We can obtain from it a reduced word for w by multiplying the corresponding letters from the right. So there exists a reduced word for w, such that we can apply a braid relation to it. A contradiction.

(c) As shown in the part (a), a 321-avoiding permutation w is defined by a sequence of numbers $i_1 < i_2 < \cdots < i_k$ and their images $w(i_1) < w(i_2) < \cdots < w(i_k)$. Draw a path from the bottom-left corner of an $n \times n$ square to its top-right corner defined as follows: the set of its "peaks", i.e. the pairs consisting of consecutive horizontal and vertical steps, pass around the boxes $(i_k, w(i_k))$. This path never passes below the diagonal. So we have a bijection between 321-avoiding permutations and paths in a square not passing below the diagonal. The number of such paths is the n-th Catalan number C_n.

11.7 A permutation w contains the pattern 2134 if and only if there exist $i < j$ and $k < m$ such that in the Rothe diagram for w the boxes (i, k) and (j, m) are free, and the boxes (i, m) and (j, k) are crossed out. It is easy to show that such numbers exist if and only if the Rothe diagram can be transformed to a Young diagram.

Chapter 12

12.1

(a) Let i be a descent in v. Then $\partial_v \mathfrak{S}_w = \partial_{vs_i} \partial_i \mathfrak{S}_w$ is nonzero only if i is a descent in w. In this case $\partial_v \mathfrak{S}_w = \partial_{vs_i} \mathfrak{S}_{ws_i}$, and the desired statement can be proven by induction on $\ell(w) = \ell(v)$.

(b) Suppose $\sum_w c_w \mathfrak{S}_w = 0$ (the sum is finite). Let v be a longest permutation such that $c_v \neq 0$. But we have $0 = \partial_v \sum_w c_w \mathfrak{S}_w = c_v$, which is a contradiction.

12.2

(a) $\mathfrak{S}_{s_i s_j} = (x_1 + \cdots + x_i)(x_1 + \cdots + x_j)$;

(b) $\mathfrak{S}_{\overline{1432}} = x_1^2 x_2 + x_1^2 x_3 + x_1 x_2^2 + x_1 x_2 x_3 + x_2^2 x_3$.

12.3 A k-Grassmannian permutation has a unique descent at k. So every divided difference except ∂_k annihilates the corresponding Schubert polynomial.

12.4

(a) $w = s_1 s_2 \ldots s_{n-2} s_{n-1} = \overline{2, 3, \ldots, n, 1}$;

(b) $w = s_{n-1} s_{n-2} \ldots s_2 s_1 = \overline{n, 1, 2, \ldots, n-1}$;

(c) $w = s_{n-k} \ldots s_{n-2} s_{n-1} = \overline{1, 2, \ldots, n-k-1, n-k+1, \ldots, n, n-k}$;

(d) $w = s_{n-1} s_{n-2} \ldots s_{n-k-1} s_{n-k}$
$= \overline{1, 2, \ldots, n-k-1, n, n-k, n-k+1, \ldots, n-1}$.

12.5 Let us compute $\mathfrak{S}_{\overline{1432}}$ using the Lascoux formula. The remaining Schubert polynomials are computed similarly.

$i = 1, j = 2,$ $\mathfrak{S}_{\overline{2134}} = x_1\mathfrak{S}_{\overline{1234}} = x_1;$

$i = 1, j = 3,$ $\mathfrak{S}_{\overline{3124}} = x_1\mathfrak{S}_{\overline{2134}} = x_1^2;$

$i = 2, j = 3,$ $\mathfrak{S}_{\overline{2314}} = x_2\mathfrak{S}_{\overline{2134}} = x_1x_2;$

$i = 2, j = 3,$ $\mathfrak{S}_{\overline{3214}} = x_2\mathfrak{S}_{\overline{3124}} = x_1^2x_2;$

$i = 2, j = 3,$ $\mathfrak{S}_{\overline{1324}} = x_2\mathfrak{S}_{\overline{1234}} + \mathfrak{S}_{\overline{2134}} = x_2 + x_1;$

$i = 2, j = 4,$ $\mathfrak{S}_{\overline{2413}} = x_2\mathfrak{S}_{\overline{2314}} + \mathfrak{S}_{\overline{3214}} = x_1^2x_2 + x_1x_2^2;$

$i = 2, j = 4,$ $\mathfrak{S}_{\overline{1423}} = x_2\mathfrak{S}_{\overline{1324}} + \mathfrak{S}_{\overline{3124}} = x_1^2 + x_1x_2 + x_2^2;$

$i = 3, j = 4,$ $\mathfrak{S}_{\overline{1432}} = x_3\mathfrak{S}_{\overline{1423}} + \mathfrak{S}_{\overline{2413}} = x_1^2x_2 + x_1^2x_3 + x_1x_2^2 + x_1x_2x_3 + x_2^2x_3.$

Chapter 13

13.1 One set is obtained from the other by reflecting it with respect to the diagonal.

13.3

(a) The length of the permutation $w = \overline{1, 2, \ldots, k-1, k+1, \ldots, n}, k$ equals $n-k$. In each pipe dream of such shape strand n goes to k and passes through all crosses vertically. These crosses are located in rows $1 \le i_1 < i_2 < \cdots < i_{n-k} \le n-1$. The sum over all possible pipe dreams gives exactly e_{n-k}.

(b) Strand k goes to n and passes through all $n-k$ crosses horizontally. These crosses are located in rows $1 \le i_1 \le i_2 \le \cdots \le i_{n-k} \le k$, hence the sum of such monomials equals h_{n-k}.

13.4 More on the relation between pipe dreams of shape $w_{1,n}$ and Catalan numbers can be found in the preprint [Woo04] by A. Woo.

(a) Use the recurrence relation for Catalan numbers:

$$C_n = \sum_{k=0}^{n-1} C_{n-k-1}C_k.$$

Given two pipe dreams $P_1 \in PD\left(w_{1,n-k-1}\right)$ and $P_2 \in PD\left(w_{1,k}\right)$, we can produce a pipe dream $P \in PD\left(w_{1,n}\right)$ as shown in Figure 16.2.
The shaded area in Fig. 16.2 shows the pipe dream P_1 (below) and two parts of the pipe dream P_2 (above). It is left to the reader to show that such a construction is well-defined and gives a bijection. It turns out that the number of pipe dreams of shape $w_{1,n}$ satisfies the recurrence relation for Catalan numbers.

(b) Define a map from the set of triangulations of an $(n+2)$-gon into the set of pipe dreams $PD\left(w_{1,n}\right)$ as follows. Index the vertices of the polygon cyclically by $1, \ldots, n+2$. For each diagonal (or side) joining i and j, with $i < j$, we draw an elbow in the corresponding pipe dream at the position $(i, n-j+3)$ (where

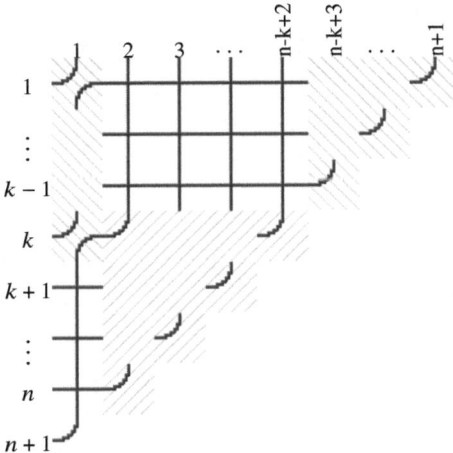

Fig. 16.2: Assembling a pipe dream $P \in \mathrm{PD}\left(w_{1,n}\right)$ out of pipe dreams $P_1 \in$ $\mathrm{PD}\left(w_{1,n-k-1}\right)$ and $P_2 \in \mathrm{PD}\left(w_{1,k}\right)$

i and $n - j + 3$ are row and column numbers, respectively). Fill the remaining positions of the pipe dream by crosses.

To show that this pipe dream indeed has the right shape, use the following idea. Any pair of triangulations can be joined by a sequence of *flips*:

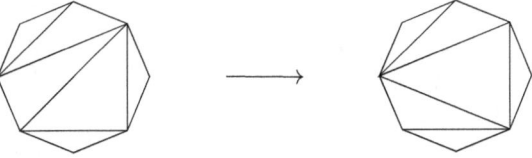

One can see that the flip of a diagonal corresponds to the following shape-preserving transform of pipe dreams:[1]

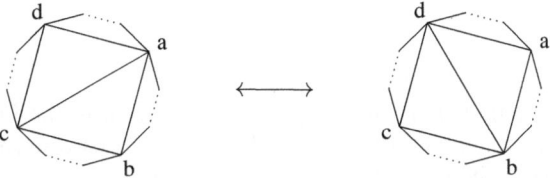

----
[1] In particular, the pipe dream graph of $w_{1,n}$, as defined in Section 14.2, is the 1-skeleton of the *Stasheff associahedron*.

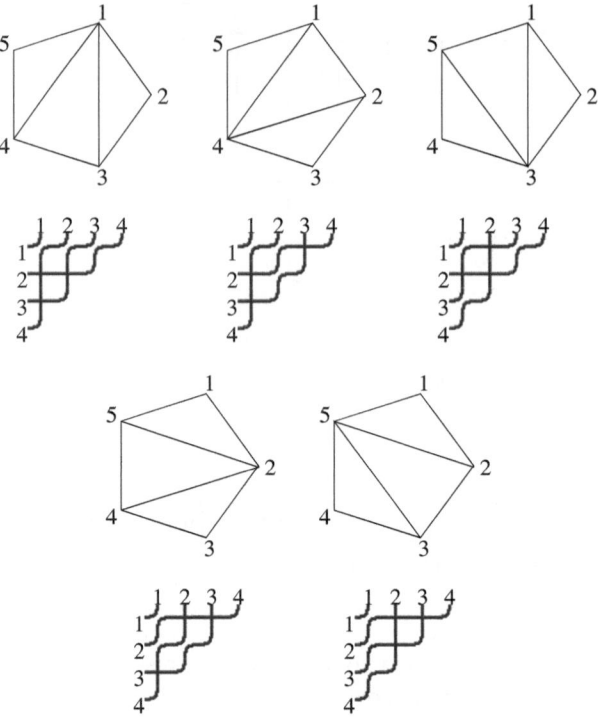

Fig. 16.3: Pipe dreams of shape $w_{1,3}$ and the corresponding triangulations of a pentagon

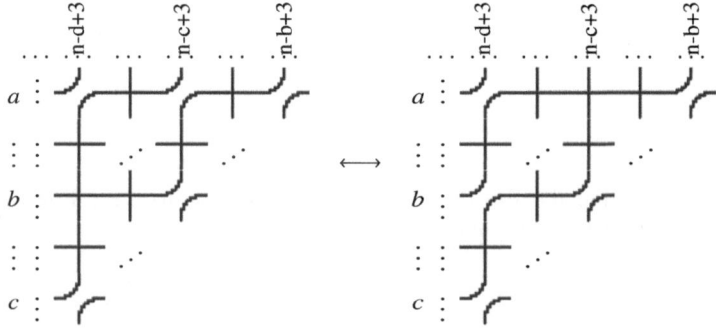

Finally, the triangulation with all diagonals adjacent to the first vertex corresponds to the pipe dream filled with elbows in the first row and crosses in the remaining ones. The shape of this pipe dream is exactly $w_{1,n}$.

This is an injective map of sets of the same cardinality, hence it is bijective.

(c) Similar to part (a).

Chapter 14

14.2 Write f as a linear combination of Schubert polynomials, act on it by ∂_i and use their linear independence.

14.3

(a) The bottom pipe dream w^{-1} is obtained by reflecting the "top" pipe dream for w with respect to the main diagonal.
(b) Its crosses form a rectangle.

14.5 (a) Reflection of a pipe dream with respect to the main diagonal corresponds to replacing each $x_i + y_j$ by $x_j + y_i$.

(b) Note that $\mathfrak{S}_{v^{-1}}(\mathbf{y})$ can be computed as follows: to each cross assign the weight y_j, where j is the number of its *column*, and take the sum over all pipe dreams of shape v. Then the formula (14.5) can be interpreted as follows. Starting from two pipe dreams, we can make a single pipe dream out of them by connecting them by elbow joints, as shown below.

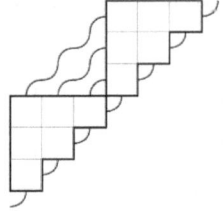

Let us put crosses and elbows to obtain a pipe dream of shape w. Suppose the weight of a cross in the i-th row of the lower part is x_i, while the weight of a cross in the j-th row of the upper part is y_j. Then the sum of monomials over all such pipe dreams equals

$$\sum_{\substack{w=vu \\ \ell(u)+\ell(v)=\ell(w)}} \mathfrak{S}_u(\mathbf{x})\mathfrak{S}_{v^{-1}}(\mathbf{y}).$$

Now consider the formula (14.5). It can be interpreted as follows: take all possible pipe dreams of shape w, and for each pipe dream consider all $2^{\ell(w)}$ possible ways to assign the weight x_i to some crosses (i being the row number) and the weight y_j, where j is the column number, to the remaining crosses. Then take the sum of all these $2^{\ell(w)} \cdot |\operatorname{PD}(w)|$ monomials.

Establish a bijection. Take a pipe dream P of shape w, with some crosses labeled by x_i, and the remaining (let us mark them by circles) labeled by y_j. Let us "pull out" these crosses to make the second pipe dream. We shall do this column by column, from right to left.

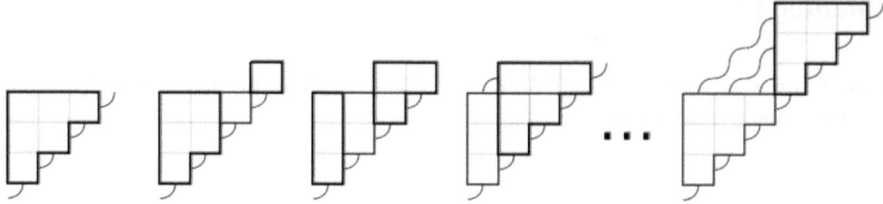

While dealing with each column, we want the following conditions to hold:

- the pipe dream shape is preserved;
- the number of circled crosses in the current column is preserved;
- the number of crosses without circles in each row is preserved.

Let us move the circled crosses one by one, starting from the top one. For each cross we distinguish between two possibilities:

- there is a cross from the right of the circled cross. In this case we can note that the circled cross is not in the bottom box of this column. Interchange the circled cross with the one without a circle (with respect to the column, the circled cross goes one box down).

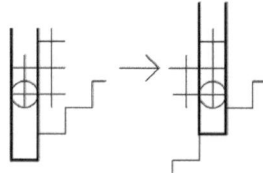

- there is an elbow joint to the right of the circled cross. In this case:

  – If above ⊕⌐ we have two elbow joints, we move the circled cross diagonally:

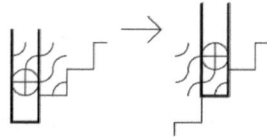

  – if above ⊕⌐ we have several pairs +⌐, then lift the whole column of crosses in the northeast direction and move the circle to the topmost cross in the column;

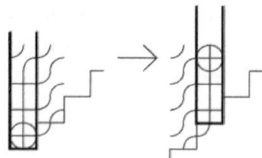

  – If above ⊕⌐ we have several pairs of crosses, then the circled cross is transferred by a ladder move.

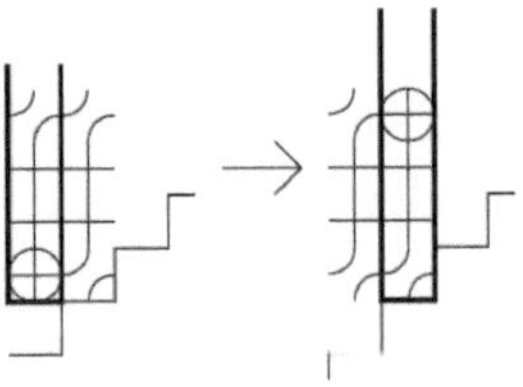

- The general case is a combination of both these situations. Above 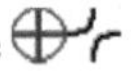 we have pairs 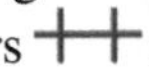 and 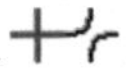 in some order, and on the top of this column we have 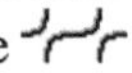 (note that 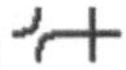 cannot occur below 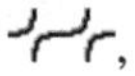, since the pipe dreams are reduced). Starting from the top, we pull the crosses from 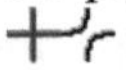 through 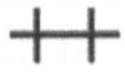 by ladder moves and mark the topmost of the transferred crosses by a circle.

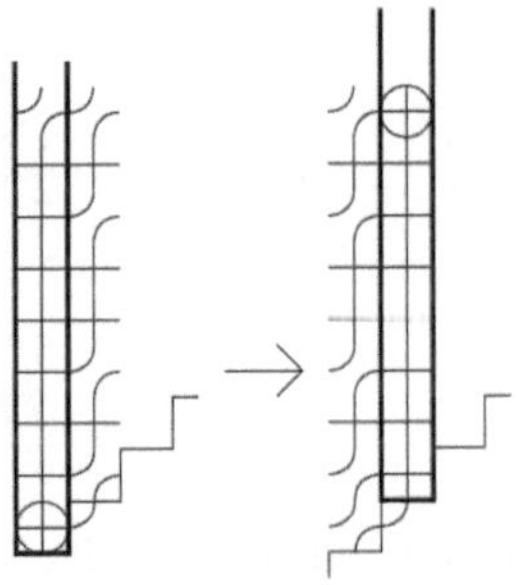

If there are several circled crosses in a column, move them starting from the topmost. In such a way we move the column northeast; applying this procedure several times, we push all circled crosses into the new pipe dream. They are all in the same columns as before. All crosses not bearing circles remain in the same rows as before. So these pipe dreams correspond to the same monomial. The last thing we need to note is that this transform is invertible. The Cauchy decomposition formula is proven.

Chapter 15

15.1

(a) Use the formula $\mathfrak{S}_w = \partial_{w^{-1}w_{0,n}} \mathfrak{S}_{w_{0,n}}$. Let Q_1 be a reduced word for $v_1^{-1}w_{0,n}$, and let Q_2 be a reduced word for $v_2^{-1}w_{0,m}$. Then $Q_1 \circ \uparrow^n (Q_2)$ is a reduced word for $(v_1 * v_2)^{-1}w_{0,n+m}$ (this is not hard to check explicitly).
Then

$$\mathfrak{S}_{v_1 * v_2} = \partial_{(v_1 * v_2)^{-1}w_{0,m+n}} \mathfrak{S}_{w_{0,m+n}} = \partial_{Q_1} \partial_{\uparrow^n(Q_2)} \mathfrak{S}_{w_{0,m+n}}.$$

Note that

$$\mathfrak{S}_{w_{0,m+n}} = x_1^{m+n-1} x_2^{m+n-2} \dots x_{m+n-2}^2 x_{m+n-1} = (x_1 x_2 \dots x_n)^m \cdot \mathfrak{S}_{w_{0,n}} \cdot \uparrow^n \left(\mathfrak{S}_{w_{0,m}}\right).$$

The operator ∂_{Q_1} acts only on $\mathfrak{S}_{w_{0,n}}$ (the remaining two factors are symmetric with respect to $x_1, \ldots, x_n$), while the operator $\partial_{\uparrow^n(Q_2)}$ acts only on $\uparrow^n\left(\mathfrak{S}_{w_{0,m}}\right)$, since the remaining two factors are symmetric with respect to $x_{n+1}, \ldots, x_{m+n}$. Hence

$$\mathfrak{S}_{v_1 * v_2} = \partial_{Q_1} \partial_{\uparrow^n(Q_2)} \mathfrak{S}_{w_{0,m+n}}$$

$$= (x_1 x_2 \ldots x_n)^m \left(\partial_{Q_1} \mathfrak{S}_{w_{0,n}}\right) \cdot \uparrow^n \left(\partial_{Q_2} \mathfrak{S}_{w_{0,m}}\right)$$

$$= (x_1 \ldots x_n)^m \mathfrak{S}_{v_1} \uparrow^n \left(\mathfrak{S}_{v_2}\right).$$

(b) Let P_1 and P_2 be pipe dreams of shape v_1, v_2 respectively. Take a rectangle of size $m \times n$ (m columns and n rows) filled by crosses, draw P_1 to the right of it and P_2 below it. We get a pipe dream of shape $v_1 * v_2$.

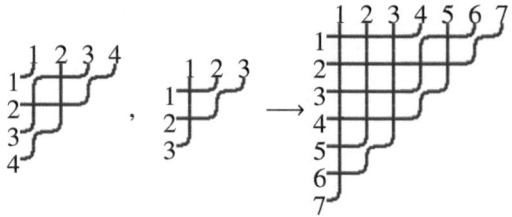

This construction is invertible: the upper left corner of each pipe dream of shape $v_1 * v_2$ contains an $m \times n$ rectangle of crosses. Removing it gives us two pipe dreams of shapes v_1 and v_2. This bijection immediately implies the desired relation.

15.2 Arrange all pipe dreams of shape w into groups with the same top row. The formula we need to prove describes this decomposition.

Suppose that the first row of a pipe dream has crosses in boxes $i_1 < i_2 < \cdots < i_k$. Then $Q_v = s_{i_k} \ldots s_{i_2} s_{i_1}$ is a descending initial word for w. Since $w(1)$ is determined only by the first row of a pipe dream, we have $w(1) = v(1)$. Removing the first row, we get a pipe dream of shape $\downarrow (v^{-1} w)$. This implies the required formula.

15.3 Act on the formula from the previous exercise by operator ∂_i:

$$\partial_i \mathfrak{B}_w = \sum_{\substack{Q_v \text{ descending initial word for } w \\ v(1)=w(1)}} x_1^{|Q_v|} \cdot \uparrow\left(\partial_{i-1} \mathfrak{B}_{\downarrow(v^{-1}w)}\right). \tag{16.1}$$

(a) Let $\ell(w s_i) > \ell(w)$, and let Q_v be a word from the right-hand side of (16.1). Then we have $\ell\left(\downarrow(v^{-1}w)s_{i-1}\right) > \ell\left(\downarrow(v^{-1}w)\right)$. Indeed, passing from w to $v^{-1}w$ corresponds to removing from a reduced word Q for w its initial subword Q_v. Since adding s_i to Q leaves it reduced, so does adding s_i to the "shortened" word.

So we have $\ell(v^{-1} w s_i) > \ell(v^{-1} w)$. Also the lowering operator does not affect the length of permutation.

Since $v^{-1}w \in S_n$, then, by the induction hypothesis, we have $\mathfrak{P}_{\downarrow(v^{-1}w)} = \mathfrak{S}_{\downarrow(v^{-1}w)}$. But then ∂_{i-1} annihilates all these polynomials, so $\partial_i \mathfrak{P}_w = 0$.

(b) Now let $\ell(ws_i) < \ell(w)$. Then any proper initial word for ws_i is also an initial word for w. Let Q_v be a word from the right-hand side of (16.1). Then $\ell(v) + \ell(v^{-1}w) = \ell(w)$.

Two cases may occur:

(i) $\ell(v^{-1}ws_i) = \ell(v^{-1}w) - 1$, and Q_v is an initial word for ws_i.
(ii) $\ell(v^{-1}ws_i) > \ell(v^{-1}w)$, and $\ell(\downarrow(v^{-1}w)s_{i-1}) > \ell(\downarrow(v^{-1}w))$.

According to the induction hypothesis, the right-hand side of (16.1) equals the sum of Schubert polynomials $\mathfrak{S}_{\downarrow(v^{-1}w)}$. Those of them that correspond to (ii) vanish after applying ∂_i. On the other hand, those corresponding to (i) become $\mathfrak{S}_{\downarrow(v^{-1}ws_i)}$. This means that

$$\partial_i \mathfrak{P}_w = \sum_{\substack{Q_v \text{ descending initial word for } ws_i \\ v(1)=ws_i(1)}} x_1^{|Q_v|} \cdot \uparrow \left(\mathfrak{S}_{\downarrow(v^{-1}ws_i)} \right)$$

$$= \sum_{\substack{Q_v \text{ descending initial word for } ws_i \\ v(1)=ws_i(1)}} x_1^{|Q_v|} \cdot \uparrow \left(\mathfrak{P}_{\downarrow(v^{-1}ws_i)} \right) = \mathfrak{P}_{ws_i}.$$

15.4

(a) Follows from the block decomposition of Schubert polynomials and the induction hypothesis.
(b) Note that every permutation $v \in S_{n+1}$, such that $v(n+1) = 1$, has the form $u * 1$. Let $w(m) = 1$ (and $m > 1$). Then $ws_m(m+1) = 1$ and $\ell(ws_m) = \ell(w) + 1$ (since m is an ascent in w). Multiplying by several consecutive letters, we get $v = ws_m s_{m+1} \ldots s_n$, with $\ell(v) = \ell(w) + n - m + 1$ and $v = u * 1$.
We have proved that $\mathfrak{S}_{u*1} = \mathfrak{P}_{u*1}$. We also know how the divided difference operators ∂_i with $i > 1$ act on $\mathfrak{S}_w$ and $\mathfrak{P}_w$. So we have

$$\mathfrak{S}_w = \partial_n \ldots \partial_{m+1} \partial_m \mathfrak{S}_{u*1} = \partial_n \ldots \partial_{m+1} \partial_m \mathfrak{P}_{u*1} = \mathfrak{P}_w.$$

15.5

(a) For $1 < i \le n$ and $\ell(ws_i) < \ell(w)$ we have

$$\partial_i (\mathfrak{S}_w - \mathfrak{P}_w) = \mathfrak{S}_{ws_i} - \mathfrak{P}_{ws_i} = 0$$

by the induction hypothesis. And if $\ell(ws_i) > \ell(w)$, we have $\partial_i (\mathfrak{S}_w - \mathfrak{P}_w) = 0$ as well.
So the difference $\mathfrak{S}_w - \mathfrak{P}_w$ is symmetric with respect to $x_2, x_3, \ldots, x_{n+1}$. But, since $w \in S_{n+1}$, these polynomials do not depend upon x_{n+1}. This means that $\mathfrak{S}_w - \mathfrak{P}_w$ is independent from $x_2, \ldots, x_n$. Since both polynomials are homogeneous of degree $\ell(w)$, we have $\mathfrak{S}_w - \mathfrak{P}_w = cx_1^{\ell(w)}$.
(b) Since $w(1) = 1 < w(2)$, we have $\partial_1 \mathfrak{S}_w = 0$. This means that the coefficients in front of $x_1^{\ell(w)}$ and $x_2^{\ell(w)}$ in $\mathfrak{S}_w$ coincide.

The coefficients in front of $x_1^{\ell(w)}$ and $x_2^{\ell(w)}$ in $\mathfrak{P}_w$ are equal; they can be either 0 or 1.

Indeed, every pipe dream of shape w has an elbow joint in its upper-left corner. If PD(w) contains a (unique) pipe dream with crosses only in the top row (it corresponds to $x_1^{\ell(w)}$), this row can be moved southwest by 1, providing a pipe dream corresponding to $x_2^{\ell(w)}$. Similarly, if all crosses in a pipe dream are in the second row, we can move them northwest to the first row.

If there are no such pipe dreams, then the coefficients in front of $x_1^{\ell(w)}$ and $x_2^{\ell(w)}$ in $\mathfrak{P}_w$ are equal to 0.

Since the coefficients in front of $x_2^{\ell(w)}$ in $\mathfrak{S}_w$ and $\mathfrak{P}_w$ coincide, so do the coefficients in front of $x_1^{\ell(w)}$. This means that $\mathfrak{S}_w = \mathfrak{P}_w$, q. e. d.

References

BB93.   Nantel Bergeron and Sara Billey. RC-graphs and Schubert polynomials. *Experiment. Math.*, 2(4):257–269, 1993.

BGG73. Joseph N. Bernšteĭn, Israel M. Gel'fand, and Sergei I. Gel'fand. Schubert cells, and the cohomology of the spaces G/P. *Uspehi Mat. Nauk*, 28(3(171)):3–26, 1973. English translation: *Russian Mathematical Surveys* 28(3):1–26, 1973.

BJS93. Sara C. Billey, William Jockusch, and Richard P. Stanley. Some combinatorial properties of Schubert polynomials. *J. Algebraic Comb.*, 2(4):345–374, November 1993.

Bre99. David M. Bressoud. *Proofs and Confirmations: The Story of the Alternating-Sign Matrix Conjecture*. Spectrum. Cambridge University Press, 1999.

Cau15. Augustin Louis Cauchy. Mémoire sur les fonctions qui ne peuvent obtenir que deux valeurs égales et de signes contraires par suite des transpositions opérées entre les variables qu'elles renferment. *Journal de l'Ecole polytechnique*, 10(17):29–112, 1815.

DK05.   Vladimir I. Danilov and Gleb A. Koshevoĭ. Massifs and the combinatorics of Young tableaux. *Uspekhi Mat. Nauk*, 60(2(362)):79–142, 2005. English translation: *Russian Mathematical Surveys* 60(2):269–334, 2005.

Egg19. Eric S. Egge. *An Introduction to Symmetric Functions and Their Combinatorics*. Student Mathematical Library. American Mathematical Society, 2019.

Eul50. Leonard Euler. Demonstratio gemina theorematis neutoniani, quo traditur relatio inter coefficientes cuiusvis aequationis algebraicae et summas potestatum radicum eiusdem. 1750.

FK96.   Sergey Fomin and Anatol N. Kirillov. The Yang–Baxter equation, symmetric functions, and Schubert polynomials. In *Proceedings of the 5th Conference on Formal Power Series and Algebraic Combinatorics (Florence, 1993)*, volume 153, pages 123–143, 1996.

Ful97. William Fulton. *Young Tableaux with applications to representation theory and geometry*, volume 35 of *London Mathematical Society Student Texts*. Cambridge University Press, Cambridge, 1997.

Gor17. Alexey L. Gorodentsev. *Algebra. II. Textbook for students of mathematics*. Springer, Cham, 2017. Originally published in Russian, 2015.

GV85.   Ira Gessel and Gérard Viennot. Binomial determinants, paths, and hook length formulae. *Advances in Mathematics*, 58(3):300–321, 1985.

Jac41. Carl G. J. Jacobi. De functionibus alternantibus earumque divisione per productum e differentiis elementorum conflatum. *Journal für die reine und angewandte Mathematik (Crelles Journal)*, 1841(22):360–371, 1841.

KM59.   Samuel Karlin and James McGregor. Coincidence probabilities. *Pacific Journal of Mathematics*, 9(4):1141 – 1164, 1959.

Knu70. Donald E. Knuth. Permutations, matrices, and generalized Young tableaux. *Pacific Journal of Mathematics*, 34(3):709 – 727, 1970.

Knu12. Allen Knutson. Schubert polynomials and symmetric functions. 2012. http://pi. math.cornell.edu/~allenk/schubnotes.pdf.

Lin73.   Bernt Lindström. On the vector representations of induced matroids. *Bulletin of The London Mathematical Society*, 5:85–90, 1973.

Lit41.   Dudley E. Littlewood. The theory of group characters. *Bull. Amer. Math. Soc*, 47:357–359, 1941.

LR34.   Dudley E. Littlewood and Archibald R. Richardson. Group Characters and Algebra. *Philosophical Transactions of the Royal Society of London Series A*, 233:99–141, January 1934.

LR35.   Dudley E. Littlewood and Archibald R. Richardson. Some special *s*-functions and *q*-series. *The Quarterly Journal of Mathematics*, os-6(1):184–198, 01 1935.

LS82.   Alain Lascoux and Marcel-Paul Schützenberger. Polynômes de Schubert. *C. R. Acad. Sci. Paris Sér. I Math.*, 294(13):447–450, 1982.

Mac91.   Ian G. Macdonald. *Notes on Schubert Polynomials*. Laboratoire de Combinatoire et d'Informatique Mathématique Montreal. Publications du LACIM, 1991.

Mac98.   Ian G. Macdonald. *Symmetric Functions and Hall Polynomials*. Oxford classic texts in the physical sciences. Clarendon Press, second edition, 1998.

Man98.   Laurent Manivel. *Fonctions symétriques, polynômes de Schubert et lieux de dégénérescence*, volume 3 of *Cours Spécialisés [Specialized Courses]*. Société Mathématique de France, Paris, 1998.

New07.   Isaac Newton. *Arithmetica universalis sive de compositione et resolutione arithmetica liber*. 1707.

Rob38.   Gilbert de Beauregard Robinson. On the representations of the symmetric group. *American Journal of Mathematics*, 60(3):745–760, 1938.

Sch01.   Issai Schur. *Über eine Klasse von Matrizen, die sich einer gegebenen Matrix zuordnen lassen*. Dieterich (Berlin), 1901.

Sch61.   Craige Schensted. Longest increasing and decreasing subsequences. *Canadian Journal of Mathematics*, 13:179–191, 1961.

Sch77.   Marcel-Paul Schützenberger. La correspondance de Robinson. In Dominique Foata, editor, *Combinatoire et Représentation du Groupe Symétrique*, pages 59–113, Berlin, Heidelberg, 1977. Springer Berlin Heidelberg.

Tho74.   Glanffrwd P. Thomas. *Baxter Algebras and Schur Functions*. University College of Swansea, 1974.

Tru64.   Nicola Trudi. Intorno un determinante piu generale di quello che suol dirsi determinante delle radici di una equazione, ed alle funzioni simmetriche complete di queste radici. *Rend. Accad. Sci. Fis. Mat. Napoli*, 3:121–134, 1864.

Vin03.   Ernest B. Vinberg. *A course in algebra*, volume 56 of *Graduate Studies in Mathematics*. American Mathematical Society, Providence, RI, 2003. Translated from the 2001 Russian original by Alexander Retakh.

Woo04.   Alexander Woo. Catalan numbers and Schubert polynomials for $w = 1(n + 1) \ldots 2$. arXiv:math.CO/0407160, 2004.

Index

© The Author(s), under exclusive license to Springer Nature Switzerland AG 2024
E. Smirnov, A. Tutubalina, *Symmetric Functions: A Beginner's Course*,
Moscow Lectures 10, https://doi.org/10.1007/978-3-031-50341-2